Current Topics in Microbiology
239 and Immunology

Editors

R.W. Compans, Atlanta/Georgia
M. Cooper, Birmingham/Alabama
J.M. Hogle, Boston/Massachusetts · Y. Ito, Kyoto
H. Koprowski, Philadelphia/Pennsylvania · F. Melchers, Basel
M. Oldstone, La Jolla/California · S. Olsnes, Oslo
M. Potter, Bethesda/Maryland · H. Saedler, Cologne
P.K. Vogt, La Jolla/California · H. Wagner, Munich

Springer-Verlag Berlin Heidelberg GmbH

Satellites and Defective Viral RNAs

Edited by Peter K. Vogt and Andrew O. Jackson

With 39 Figures and 6 Tables

Springer

Professor PETER K. VOGT, Ph.D.

Division of Oncovirology,
Mail Drop & Room BCC239
Department of Molecular and Experimental Medicine
The Scripps Research Institute
10550 N. Torrey Pines Road
La Jolla, CA 92037
USA

Andrew O. Jackson, Ph.D.

University of California, Berkeley
Department of Plant and Microbial Biology
Berkeley, CA 94720
USA

Cover Illustration: Models proposed for in vitro structure of Q-sat-RNA and Ix5-sat-RNA (modified from Fig. 2A, B, Garcìa-Arenal and Palukaitis, this volume)

Cover Design: Design & Production GmbH, Heidelberg

ISBN 978-3-662-09798-4   ISBN 978-3-662-09796-0 (eBook)
DOI 10.1007/978-3-662-09796-0

© Springer-Verlag Berlin Heidelberg 1999
Library of Congress Catalog Card Number 15-12910
Originally published by Springer-Verlag Berlin Heidelberg New York in 1999.
Softcover reprint of the hardcover 1st edition 1999

Typesetting: Scientific Publishing Services (P) Ltd, Madras

Production Editor: Angélique Gcouta

SPIN: 10684979   27/3020 - 5 4 3 2 1 0   Printed on acid-free paper

Preface

The nine chapters presented in this book provide contemporary reviews of research on defective RNAs, satellite RNA viruses, and dependent RNA viruses that require the presence of a helper virus in order to establish productive infections. Since their initial identification nearly four decades ago, fundamental analyses of pathogenic and interdependent interactions involving these agents have contributed enormously to our appreciation of virus structure, RNA replication, and processes leading to disease. Findings arising from these studies have also advanced numerous ancillary areas, including structure and function of nucleic acids and proteins, nucleoprotein interactions, translational mechanisms, RNA processing, macromolecular evolution, and a plethora of other specialty topics. Research on these subviral pathogens is continuing to illuminate various aspects of biology, chemistry, and biotechnology, so the book is intended to provide a current treatment that will be useful for readers with interests in topics related to these areas.

Observations in the early 1960s first revealed that a defective virus, satellite tobacco necrosis virus (STNV), is associated with and depends on the presence of a helper virus, tobacco necrosis virus (TNV), for its multiplication. This finding ushered in a new era in virology that soon resulted in a more profound appreciation of the relationships of viruses and their interactions with each other. The demonstration that both STNV and TNV are transmitted by a soil fungus and that STNV is a molecular parasite that attenuates symptoms elicited by the helper virus also stimulated considerable interest in the role of defective agents in virus biology and pathogenesis. Thus, within a decade, attempts to understand the interdependence of small defective agents led to the discovery of multicomponent viruses, viroids, and satellite RNAs (satRNAs) that depend on helper viruses for replication and encapsidation. Several different classes of satRNAs, satellite-like agents, and codependent viruses are currently known to be associated with plant RNA viruses and to affect a large number of disease syndromes. In contrast, defective RNAs that evolve as a consequence of errors in viral genome replication are relatively rare among plant viruses, even though this phenomenon is fairly ubiquitous in animal virus systems. On the other hand, satellite-like RNAs have not become widely distributed among the commonly studied animal viruses: The sole example on the biomedical front is hepatitis delta virus (HDV), which is supported by hepatitis B virus. The circular satRNA genome of HDV has structural similarities to viroids and also encodes a protein, the delta antigen, which may have evolved by capture of a host gene. Thus, HDV provides the first demonstration that a satRNA similar to those found in plants is a major component of an animal virus disease complex. We believe it is likely that additional defective agents that are more refractory to discovery than the present example may coexist with animal viruses, so hopefully the reviews presented in this series will stimulate interest in this possibility as well as encourage others to initiate research into related interactions that affect plant viruses.

The book begins with a chapter by Andrew White and Jack Morris focusing on deletion, recombination, and selection events leading to evolution of defective subviral RNAs from the genomes of plant monopartite plus-strand RNA viruses. Analysis of these processes in plant systems provides some notable advantages over studies in animals because the complicating effects of the immune system can be avoided. Therefore, more direct interpretations can be made of the interactions of defective RNAs with their parental helper viruses, their plant hosts, and the environment that lead to variable disease attenuation. The authors discuss these variables and suggest that if they can be manipulated appropriately, defective RNAs may have considerable future potential for disease control and possible use as vectors for foreign gene expression.

Linear satRNAs and various aspects of their evolution, helper virus dependence, and roles in disease development are discussed in the next three chapters. SatRNAs were classically defined as defective RNAs lacking appreciable sequence relatedness to the helper virus genomes that support their replication and provide coat protein for encapsidation of their RNA. Yet, subsets of satellite-like derivatives deviate in substantial ways from this definition. Among these are a few chimeric derivatives in which a portion of the sequence of the dependent agent clearly is derived from the helper virus and a portion consists of sequences with little obvious relatedness to the helper virus. The most carefully studied molecules of this nature are associated with turnip crinkle virus (TCV), which harbors a novel family of satRNAs with variable structure and symptom modulating effects. Anne Simon provides an extensive discussion of these satellites in the second chapter and describes TCV satellite-helper relationships and host interactions that affect the disease phenotype. However, the primary focus of her chapter revolves around *cis* and *trans* factors affecting replication and the recombination and repair events involved in maintenance and evolution of the satellite derivatives. She also points out that modern plant genetic tools are now sufficiently advanced that they can be used to identify host components that are involved in the replication and pathology of these satellites.

In the third chapter, Fernando Garcia-Arenal and Peter Palukaitis provide a detailed review of a distinct class of linear satRNAs found in association with isolates of cucumber mosaic virus (CMV). These well-characterized interactions have considerable practical significance because CMV is one of the most widespread and destructive plant viruses known, and some satRNA interactions contribute to a lethal necrosis that can cause serious disease losses in crop species throughout the world. The authors summarize the discovery of CMV satRNAs, the relatedness of different strains, and the genetic determinants that contribute to exacerbation or attenuation of the disease phenotype. They also emphasize the importance of satRNA–helper virus interactions with their plant hosts and discuss the pros and cons for the use of satRNAs for disease control.

The TCV and CMV satRNA genomes are relatively small, with respective sizes that vary from slightly less than 200 to slightly more than 400 nt. However, the nepoviruses, bamboo mosaic virus (BaMV), some umbraviruses, and beet necrotic yellow vein virus (BNYVV) support more diverse complexes of satRNAs and satellite-like RNAs that range in size from 0.7 to nearly 2 kB. The chapter by Mike Mayo, Michael Taliansky and C. Fritsch illustrates the structural variability exhibited by these "larger" satellites and the range of parasitic to mutualistic effects exerted on their helper viruses. Unlike the smaller satRNAs discussed in the chapters by Simon and Garcia-Arenal and Palukaitis, a subset of the larger satRNAs associated with the nepoviruses and potexviruses have mRNA activity, and some evidence suggests that the encoded proteins may provide *trans*-acting factors needed for replication of the satRNAs. In contrast, the encoded protein of a satRNA supported by the potexvirus BaMV appears not to be essential for replication, even though the protein is found in infected plants. The authors also describe several extremely interesting interactions that require participation of the larger satRNAs. One of these involves a mutualistic umbravirus disease complex (see the chaper by Falk et al. for a comprehensive discussion of these complexes) associated with a satellite-like RNA. This complex depends on a helper virus, groundnut rosette virus (GRV), for replication, plus an association with a luteovirus helper. The involvement of the GRV satellite-like RNA in the complex is crucial because this RNA is absolutely essential for disease development, and also provides a factor necessary for the obligatory aphid transmission required for GRV survival in nature. Another interesting mutualistic complex that is described relates to the participation of BNYVV RNAs in rhizomania disease. BNYVV isolates recovered from beet roots exhibiting severe rhizomania in the field contain up to five RNA components that can be transmitted by a primitive chytrid fungus. RNAs 1 and 2 are infectious when coinoculated manually to experimental hosts, but are unable to move into roots and also lose the ability to be transmitted through the soil by the fungus. The three smaller RNAs are defective and depend on RNAs 1 and 2 for replication. However, these ancillary RNAs are important components in the rhizomania complex. Their individual synergistic functions contribute to virus transport to the roots, the efficiency of fungal transmission, and exacerbation of the disease phenotype. Clearly, each of the parasitic and mutualistic systems outlined in the chapter present a large number of fundamental problems whose resolution over the next few years will provide major contributions to virus biology, plant pathology, and the ecology of codependent virus complexes.

The next two chapters describe satellites with genomes that have similarities to viroids. The small virusoids reviewed by Bob Symons and John Randles constitute a class of satRNAs that, unlike viroids, are unable to infect plants autonomously and require a helper virus for replication and encapsidation of the circular RNA genome. The authors initially outline the identification and biological properties of the five known virusoids and focus on comparisons of the sequence relatedness and highly base-paired structures of their genomes. They then highlight differences in the mechanism of replication of different virusoids and viroids that have been deduced from variations in cleavage and circularization during RNA amplification. Symons and Randles conclude the review by discussing the information content of the tiny virusoid genomes, and they also raise several outstanding questions about genome function that need to be answered in future experiments. These include identification of polymerase promoter elements, characterization of signals mediating encapsidation by the helper virus capsid protein, and genome interactions affecting host metabolism.

The next chapter shifts the emphasis from plants to HDV, the only satRNA known to have a major involvement in disease development in animals. It has been only slightly more than a decade since the discovery that the "delta" agent supported by hepatitis B has a circular single-stranded RNA genome with some similarities to plant viroids. Since that time, HDV comparisons with viroids and virusoids have resulted in remarkable progress that has led to a better understanding of the complexity, evolution, and replication of the HDV genome, as well as its role in exacerbation of the severity of hepatitis and hepatocarcinoma. The chapter by John Taylor concentrates on current studies of HDV replication in transfected cell lines and provides a detailed overview of experimental approaches and problems in dissecting transcription of delta antigen mRNAs and characterizing promoters on the genomic RNA. Prominent among the outstanding issues are exactly how and where transcription is initiated and the mechanisms that are utilized to achieve a balance between transcription, polyadenylation, and replication. The editing process involved in production of mRNAs directing synthesis of the small and large delta antigens is reviewed, and the probable roles of the antigens in regulating the replication cycle are discussed. Taylor summarizes by predicting that some major surprises may yet arise as the mechanisms of genome replication are dissected. He reiterates the advances made by testing analogies to plant viroid and satellite systems and suggests that extrapolations to a broader range of nucleic acid replicons might also provide additional profitable insights into genome replication.

The first satellite virus, STNV, was identified in the early 1960s, and since that time, studies of these agents have contributed to resolution of several basic biological problems. However, despite their emphasis in the scientific literature, satellite viruses are quite rare, as four decades of research have revealed only three examples in addition to STNV. Monopartite spherical viruses support the replication of three satellite viruses, STNV, satellite panicum mosaic virus (SPMV) and satellite maize white line mosaic virus (SMWLMV), which are reviewed in the chapter by Karen Scholthof, Rick Jones and Andy Jackson. The SPMV and SMWLMV systems, like STNV, have restricted host ranges, and both SPMV and SMWLMV aggravate pasture and field disease problems. Allen Dodds provides an overview of satellite tobacco mosaic virus (STMV), which is unique in being supported by a rod-shaped virus, the tobacco mild green mosaic strain of TMV. STMV is restricted in nature to a native tree tobacco and appears to have evolved as a reasonably innocuous complex that poses no obvious agricultural problems. Taken together, the evidence presented in these two chapters shows that individual satellite viruses exhibit some minor strain and sequence heterogeneity, and the sequence data suggest that the satellite viruses coevolved with their helper viruses rather than originating from a common progenitor. Nevertheless, satellite viruses do share a number of common properties in their virus architecture, RNA structure, and mechanisms involved in gene expression. Both chapters highlight the extraordinary impact that studies of these model systems have had on virus biology, pathology and biochemistry, and the authors predict that future advances will focus on the general areas of evolution, genetic diversity, parasitic interactions and biotechnology.

Bryce Falk, Tongyan Tian and Hsin-Hung Yeh conclude the series by considering codependent virus systems in which luteovirus serve as helper factors for defective replicons that are able to replicate in single cells, but lack one or more functions that are essential for survival alone. Many of these complexes shift the symptom phenotype to produce a distinct disease syndrome that is easily differentiated from that induced by the luteovirus helper alone, and some appear not to alter symptom development. A common feature of all of these systems involves transencapsidation of the defective RNA within the luteovirus capsid protein to mediate aphid transmission of the dependent member of the complex. However, the systems are quite diverse: some have the ability to establish systemic infections and some do not. Thus, the dependent agent may depend on the luteovirus helper for movement functions as well as vector transmission. Throughout the chapter, Falk and colleagues focus on a discussion of the biology, interactions and genome organization of three classes of luteovirus helper dependence. The first complexes discussed are the umbravirus systems that depend on a luteovirus helper for aphid transmission. These viruses are able to move systemically, but lack the ability to synthesize a capsid protein. In these cases, luteovirus-specific aphid dispersal and, hence, survival in nature, are gained following transencapsidation of the luteovirus RNA by the luteovirus coat protein. A different mutual codependent complex is formed by a bipartite virus, pea enation mosaic virus (PEMV) whose aphid transmission has been studied for many years. RNAs 1 and 2 forming the complex both encode the replicase proteins needed for their individual amplification in single cells, but both are required for productive PEMV virion formation, for systemic movement of the virions, and for aphid transmission. The authors also discuss a third class of dependence involving luteovirus–subviral RNA interactions. The best characterized example is beet western yellows luteovirus (BWYV) and an associated RNA designated ST9, which accentuates the BWYV disease phenotype dramatically. ST9 is able to multiply in inoculated cells, but does not encode a capsid protein and also is unable to move to adjacent cells. However, ST9 enhances the accumulation of BWYV substantially in coinfected cells, which apparently provides increased pools of the BWYV movement protein needed to permit systemic movement of ST9 and increased levels of coat protein needed to produce transencapsidated ST9 virions for aphid transmission. The authors speculate that the ability of ST9 to amplify BWYV replication also contributes to the survival of BWYV, so that both partners benefit from the interaction. This hypothesis represents only one of the many issues affecting these codependent systems that remain to be elaborated on in future experiments.

We would like to thank the contributors for their patience during the assembly of this volume and for the time and energy they have invested to ensure that their chapters represent a current overview of their topics. Those readers who are unfamiliar with the area will find that the subject matter is diverse and that the authors have largely avoided overlap between related areas. Hopefully, all readers will gain insight from these contributions that will enhance their individual research and teaching activities.

A.O. Jackson and P.K. Vogt

List of Contents

List of Contributors

(Their addresses can be found at the beginning of their respective chapters.)

DODDS, J.A.   145

FALK, B.W.   159

FRITSCH, C.   65

GARCÍA-ARENAL, F.   37

JACKSON, A.O.   123

JONES, R.W.   123

MAYO, M.A.   65

MORRIS, T.J.   1

PALUKAITIS, P.   37

RANDLES, J.W.   81

SCHOLTHOF, K.-B.G.   123

SIMON, A.E.   19

SYMONS, R.H.   81

TALIANSKY, M.E.   65

TAYLOR, J.M.   107

TIAN, T.   159

WHITE, K.A.   1

YEH, H.-H.   159

Defective and Defective Interfering RNAs of Monopartite Plus-strand RNA Plant Viruses

K.A. White[1] and T.J. Morris[2]

1 Introduction

The defective interfering (DI) RNAs represent one of several classes of symptom-modulating RNAs identified in association with RNA plant virus infections. Structurally, these molecules are derived from, and represent mutant forms of, the viral genome (Perrault 1981; Lazzarini et al. 1981). DI RNAs may contain distinct types of modifications; however, the most prevalent is the deletion of one or

[1] Department of Biology, York University, 4700 Keele Street, Toronto, Ontario, M3 J 1P3, Canada
[2] School of Biological Sciences, University of Nebraska, Lincoln, NE 68588-0118, USA
KAW is supported by grants from the Natural Sciences and Engineering Research Council of Canada. TJM is supported in part by DOE grant no. DE-FG02 91ER20026.

more large segments of sequence. Despite structural differences, the common effect of the mutation(s) is to render the DI RNAs dependent on their nondefective 'parental' genome for essential viral replication proteins. This, in turn, limits replication of these molecules to cells which are coinfected with the parental genome.

The genetic relatedness between DI RNAs and their helper viruses is what distinguishes this class of molecule from another common plant virus-associated group, the satellite RNAs (ROOSSINCK et al. 1992). Although satellite RNAs are also dependent on a helper virus for replication, unlike DI RNAs, they do not share any significant nucleotide sequence identity with the genome of the virus with which they associate. When DI RNAs are present in a plant virus infection, they may alter significantly both helper virus titer and the symptoms normally induced (HILLMAN et al. 1987; JONES et al. 1990; LI et al. 1989). In other cases, these virally derived RNAs do not notably interfere with either virus accumulation or symptoms and, as a consequence, are referred to as D RNAs (WHITE et al. 1991; MAWASSI et al. 1995a).

The first definitive plant virus DI RNA to be described was found associated with the small positive-sense RNA icosahedral virus, tomato bushy stunt virus (TBSV; HILLMAN et al. 1985, 1987). Prior to this discovery, DI-like particles had been described for negative-strand plant rhabdo-like and bunya-like viruses (ADAM et al. 1983; VERKLEIJ and PETERS 1983), but it was uncertain whether positive-strand plant viruses possessed the ability to generate and/or sustain DI RNAs. Subsequent to the molecular proof for the presence of DI RNAs in TBSV, additional reports emerged which further supported the existence of similar plant virus-associated replicons (LI et al. 1989; BURGYAN et al. 1989; ROCHON 1991). Other studies then went on to show that these molecules are produced de novo from the parental genome during infections (LI et al. 1989; KNORR et al. 1991; BURGYAN et al. 1991; ROCHON 1991). To date, DI RNAs have been definitively identified with RNA plant viruses containing double-stranded, ambisense, and positive-sense genomes, and a review of DI RNAs associated with all three of these genome types in multipartite viruses was recently published (GRAVES et al. 1996). This review focuses on defective RNAs of monopartite RNA plant viruses and specifically on D and DI RNAs associated with plus-sense RNA plant viruses.

2 Structure

The DI RNAs associated with monopartite plus-strand RNA plant viruses exhibit a variety of structures. One unifying element among all characterized so far is that they contain at least one large deletion. In terms of size, these molecules are generally 10–25% the length of the parental genome. Interestingly, although considerable structural variation can be observed in DI RNAs derived from the same virus, certain highly conserved segments seem to be strictly conserved within these varying structural contexts. The invariant segments presumably correspond to sequences which are important or essential for DI RNA viability.

2.1 Single Deletions and Coding Capacity

D RNAs associated with the rod-shaped *Potexvirus* and *Closterovirus* genera contain single large deletions of a central portion of their genomes (Table 1). The positive-strand genomes of the closteroviruses are the largest among RNA plant viruses (Dolja et al. 1994) and several sizes of D RNAs have been found associated with the 19.3-kb citrus tristeza closterovirus (CTV) genome (Mawassi et al. 1995a,b). These molecules vary from 2.4 to 4.5 kb in length, but all contain the 5′- and 3′-terminal regions of the genome. Initially, the positions of the junction sites in these molecules appeared to be random. More recently, a number of independently isolated D RNAs have been identified which contain downstream deletion sites that map to a specific nucleotide position in the genome (Yang et al. 1997b), and this observation has led to the proposal of a possible mechanism for the formation of these molecules (see Sect. 3.3). Depending on their size and structure, CTV D RNAs may encode complete open reading frames (ORFs) derived from the 3′ terminus of the genome (Fig. 1a). Presently, it is unclear as to whether the products encoded in these molecules are expressed, but the recent construction of an infectious clone of a CTV D RNA should help to answer this and other questions (Yang et al. 1997a).

D RNAs have also been identified in the flexible rod-shaped potexviruses clover yellow mosaic virus (ClYMV; White et al. 1991) and cassava common mosaic virus (CsCMV; Calvert et al. 1996; Table 1). In both cases, single internal deletions resulted in the removal of large segments of the 6- to 7-kb RNA genomes. An intriguing feature of all of the D RNAs analyzed was that the deletion invariably led to the in-frame fusion of the N- and C-terminal regions, respectively, of the first and last ORFs encoded in the genome (Fig. 1b). The presence of these fusion ORFs in D RNAs with different junction sites suggested a possible selective advantage to those molecules maintaining a continuous coding region. Additional studies performed on a prototypical D RNA of ClYMV showed a positive correlation between maintenance of the fusion ORF and D RNA viability in whole plants (White et al. 1992b). Further data suggested that translatability, rather than production of the encoded product, is the important property for progeny accumulation (White et al. 1992a).

Table 1. Monopartite plus-strand RNA plant viruses found associated with D or DI RNAs

Genus	Virus	Defective RNA type (D or DI)
Closterovirus	CTV	D RNA
Potexvirus	ClYMV	D RNA
	CsCMV	D RNA
Carmovirus	TCV	DI RNA
Tombusvirus	CIRV	DI RNA
	CNV	DI RNA
	CyRSV	DI RNA
	TBSV	DI RNA

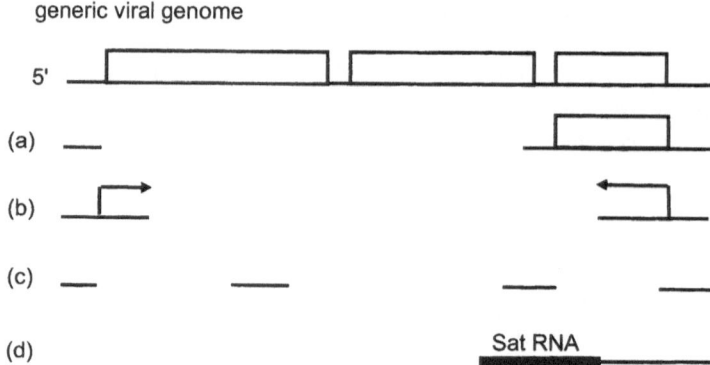

Fig. 1a–d. Examples of structural variants of D and DI RNAs associated with monopartite plus-strand RNA plant viruses. A generic viral genome is shown at the *top* with encoded ORFs shown as boxes. *a* 5′-3′ D RNA, encoding a complete viral ORF; *b* 5′-3′ D RNA, encoding a fusion ORF; *c* mosaic type DI RNA, containing internal regions of the genome; *d* hybrid DI RNA, containing both satellite and viral sequences

2.2 Multiple Deletions and Other Modifications

Members of the family *Tombusviridae* are isosahedral viruses, which include the *Tombusvirus* and *Carmovirus* genera (Russo et al. 1994; Table 1). These viruses contain 4- to 5-kb plus-sense RNA genomes, and their associated DI RNA species tend to contain several large deletions and no significant continuous coding regions. Common structural features have been identified in the DI RNAs associated with various *Tombusviruses* [TBSV, Knorr et al. 1991; cucumber necrosis virus (CNV), Finnen and Rochon 1993; cymbidium ringspot virus (CyRSV), Burgyan et al. 1991; carnation Italian ringspot virus (CIRV), Rubino et al. 1995; Table 1]. These molecules maintain noncontiguous segments corresponding to terminal regions of their genomes, as well as two internal segments (Fig. 1c). In addition to these large deletions, other more minor modifications such as segment duplication, additional smaller deletions, and nucleotide insertions, deletions, or substitutions, have been observed (Knorr et al. 1991; Burgyan et al. 1991; White and Morris 1994a).

The DI RNAs associated with the carmovirus turnip crinkle virus (TCV; Table 1) are composed of several noncontiguous segments (Li et al. 1989). An interesting structural feature observed in a TCV DI RNA, termed DI RNA G, is an insertion of 12 nonviral nucleotides near the 5′ terminus. Although the origin of these residues is unclear, it is possible that they are the result of either divergent evolution of TCV sequence, recombination with a satellite RNA, or introduction of a nonviral segment (Li et al. 1989).

2.3 Hybrid Structures

In addition to the typical DI RNAs which have been found associated with TCV, a most intriguing hybrid form of virus-associated molecule has also been described

(SIMON and HOWELL 1986). This molecule possesses sequences common to both the parental genome and a satellite RNA found to be associated with TCV (Fig. 1d). The 3' half of this RNA shows significant sequence identity to the 3' terminus of the genome, whereas the 5' half contains a sequence common to a TCV satellite RNA. The unusual 5'-satellite/3'-DI RNA structure of this molecule has made the classification this sat-RNA C molecule somewhat ambiguous. The structure of this unique molecule suggests that it was formed by a recombination event between an authentic sat-RNA and the 3' end of the TCV genome (SIMON and HOWELL 1986).

3 Formation and Accumulation

DI RNAs do not normally accumulate to readily detectable levels in natural infections of plants (CELIX et al. 1997), although they have been detected in field isolates (MAWASSI et al. 1995a,b). Generally, these molecules tend to accumulate significantly only after successive passaging at artificially high inoculum concentrations. It is believed that these conditions favor the coinfection of cells with both helper and newly formed DI RNAs, which in turn promotes further amplification. However, using the sensitive technique of reverse transcriptase-polymerase chain reaction, TBSV DI RNAs have been detected in artificially infected plants without passage at high inoculum concentrations (LAW and MORRIS 1994). This observation suggests that the high multiplicity of infection passages is not essential for the formation of DI RNA, but instead likely helps to amplify low levels of existing molecules. Interestingly, a mutant form of the CNV genome, which is unable to express the nonstructural 20-kDa protein, is able to rapidly generate significant levels of DI RNAs without serial passage (ROCHON 1991).

It is unclear why DI RNAs do not generally accumulate to high levels in natural infections, but certain viral mechanisms have likely evolved that act to suppress the formation and/or accumulation of defective genomes. It has been suggested that the CNV 20-kDa protein may act in this capacity (ROCHON 1991). Other factors which may influence DI RNA accumulation in natural infections such as environmental conditions (e.g. temperature) have been investigated, but not extensively (JONES et al. 1990; INOUE-NAGATA et al. 1997). There are, however, several examples of the influence that the host can have on the formation and/or accumulation of these molecules. For example, by passaging a viral inoculum containing DI RNAs through certain hosts, the structure of the dominant population of DI RNAs may be changed (CHANG et al. 1995) or the DI RNAs may no longer be encapsidated (ROMERO et al. 1993). Host effects may also help to explain why no associated DI RNAs have been observed with one apparent member of the *Tombusviridae* family (RUBINO and RUSSO 1997); that is, a DI RNA-permissive host may not yet have been identified. As mentioned previously, viral-specific factors may also limit the formation or accumulation of these molecules; however, genome size does not appear to be limiting, since DI and D RNAs have been found associated with both the smallest (carmoviruses) and largest (closteroviruses) monopartite plus-sense RNA viruses (LI et al. 1989; MAWASSI et al. 1995a,b).

At present, the most widely accepted model for the generation of D and DI RNAs is one which is linked to the process of genome replication (PERRAULT 1981; LAZZARINI et al. 1981), but alternative mechanisms not involving the viral polymerase could also be contributing to their generation (NAGY and SIMON 1997). Regardless of the mechanism by which these molecules are generated, it is likely that multiple structural forms are produced initially, but only those molecules which replicate efficiently are able to accumulate.

3.1 Models for Formation

Currently, the most widely accepted model for RNA virus recombination and D or DI RNA formation is the copy-choice mechanism (PERRAULT 1981; LAZZARINI et al. 1991). In this model, recombination is mediated by an actively copying viral polymerase dissociating from its template, along with its nascent strand, and re-initiating synthesis at a new position on the same template or on a different template. The sequences which are not copied are thus deleted from the newly synthesized strand. Although definitive proof of this mechanism is still lacking, there is mounting circumstantial evidence for a role for the viral replicase in the generation of deletions and other modifications found in D and DI RNAs. For example, promoter-like sequences have been found at the junctions of deletions (CASCONE et al. 1990), and their presence suggests a possible role in reinitiation of a dissociated polymerase. Nontemplated nucleotides have also been identified at junction sites, and it has been postulated that the viral replicase may be responsible for adding these residues during deletion events (LAW and MORRIS 1994; NAGY and BUJARSKI 1995). Although the viral replicase is likely involved in most recombination events involving viral sequences, other mechanisms cannot be ruled out (NAGY and SIMON 1997). For instance, a recent report on RNA recombination of QB phage suggests that transesterification may be involved in the joining of noncontiguous RNA sequences (CHETVERIN et al. 1997).

3.2 Role of Viral Polymerase

If formation of D and DI RNAs is mediated primarily by the viral polymerase, one may predict that the replicase associated with such viruses may have special properties (PERRAULT 1981; LAZZARINI et al. 1981; NAGY and SIMON 1997). For example, they may have a tendency to be nonprocessive, so as to easily dissociate from a template. In this regard, the virus-encoded polymerase components of the *Tombusvirus* genus do not possess any identifiable helicase activity, and this feature may make these viruses more susceptible to polymerase dissociation (KOONIN and DOLJA 1993; WHITE 1996). More direct evidence for polymerase involvement in recombination comes from studies in other systems. Analysis of the 1a replication component of BMV has shown that certain modifications in this protein can inactivate the ability of the viruse to undergo nonhomologous recombination (FIGLEROWICZ et al. 1997). An extract from TCV-infected plants that is capable of catalyzing viral RNA synthesis has also been shown to mediate viral RNA recombination (SONG and SIMON 1994, 1995; NAGY et al. 1997). Definitive evidence that plant virus replicases can mediate RNA recombination will require that such activity be demonstrated in vitro with highly purified replicase components.

3.3 Role of Template Structure

RNA sequence and structure is thought to influence deletion and recombination events associated with D and DI RNA formation. The analysis of junction sites, along with functional studies, has led to the suggestions that the following structural features of RNA may influence the selection of deletion sites: (a) sequence identity, (b) secondary structure, (c) promoter-like elements, and (d) 5' termini (Fig. 2). It is thought that sequence identity may aid replicase reinitiation through base pairing between the polymerase-bound nascent strand and an upstream segment of the template (Fig. 2a). In support of this idea, short stretches of sequence identity have been observed at the junction sites in different DI RNAs (KNORR et al. 1991; WHITE and MORRIS 1994b). Studies carried out on the formation of the deletions in TBSV precursor DI RNAs suggested a role of sequence identity (WHITE and MORRIS 1995); however, recent results indicate that sequence identity does not appear to significantly promote replicase-mediated deletions (Wu and White, unpublished data). Additional studies will be required to resolve this apparent discrepancy.

The evidence for a role of secondary structure in *Tombusvirus* DI RNA formation is more convincing. Two separate studies have suggested that the presence of secondary structure in the viral RNA will target deletion sites to the base of the structure (Fig. 2b; WHITE and MORRIS 1995; HAVELDA et al. 1997). The proposed mechanism is akin to that suggested for heterologous recombination in BMV (NAGY and BUJARSKI 1993), where, when challenged with a strongly base-paired region, the polymerase is sometimes able to traverse the base of the stem structure and continue synthesis.

For DI RNAs of TCV, promoter-like sequences have been identified at junction sites (CASCONE et al. 1990), and it was suggested that these elements may be involved in the reinitiation of the dissociated polymerase during plus-strand synthesis (Fig. 2c). There is now considerable experimental evidence supporting a role of various promoter-like sequences in recombination events in TCV (CASCONE et al. 1993; CARPENTER et al. 1995).

More recently, a new model was proposed to explain the formation of one class of CTV D RNA (YANG et al. 1997b). It was observed that a number of smaller D RNAs had 3' segments which corresponded precisely to one of the viral subgenomic (sg) RNAs. Based on the structures of various D RNAs, it was suggested that they were likely formed by synthesis of a negative strand complementary to the sgRNA,

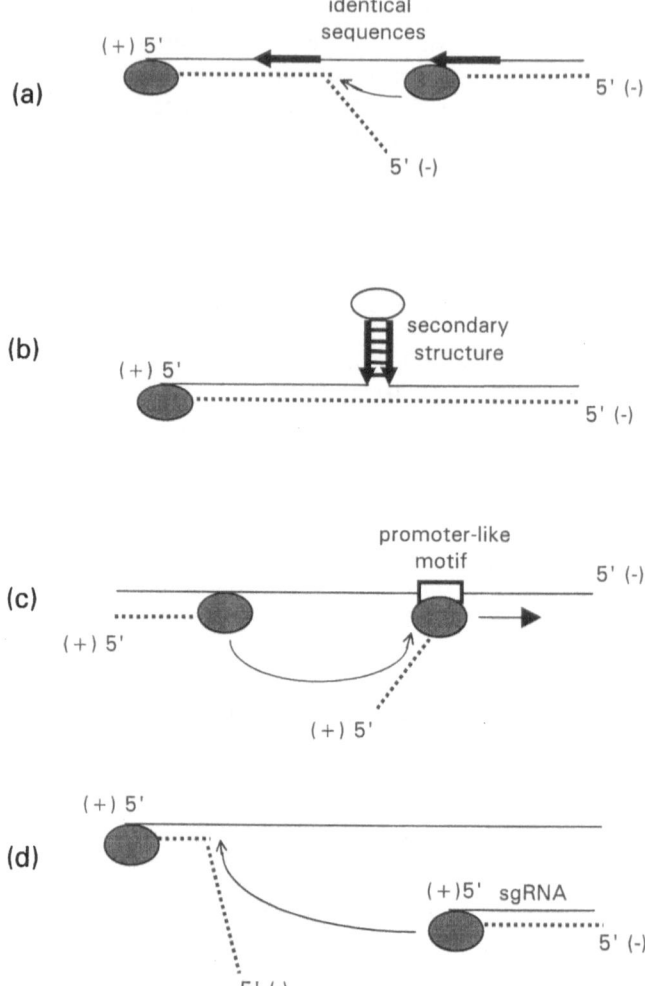

Fig. 2a–d. Schematic representation of possible mechanisms for the formation of D and DI RNAs. Viral genomes are depicted as *solid lines*, while complementary nascent RNA strands are shown as *dashed lines*. The polarity of the RNA strands is indicated [positive (+), negative (−)]. The viral replicase is represented by *shaded ovals* and its path is indicated by *thin arrows*. The deletion or recombination may be mediated by *a* identical sequences (shown as *boldface arrows*), *b* secondary structure, *c* promoter-like motifs (shown as *a box*), or *d* 5′ termini (that of the sgRNA). See text for details

followed by a template switch to the genomic RNA (Fig. 2d). This recombination mechanism of running off the end of one template and reinitiating on another template has also been proposed as a mechanism of *trans*-recombination in TBSV (WHITE and MORRIS 1995).

It is clear from this survey that the role of RNA sequence and structure in D and DI RNA formation may be varied and that different mechanisms are likely to be involved. It is also possible that other structural properties such as sequence composition, as has been demonstrated for recombination events in BMV (NAGY and BUJARSKI 1997), may influence these deletion events. More detailed studies on template properties as they relate to deletion events are therefore required to fully understand the process(es) of D and DI RNA formation.

There are limitations to elucidating the mechanisms operating in the formation of DI RNAs through the analysis of sequences present at junction sites. It is important to keep in mind the role of selection, because only those molecules which are replicable will be competitive and therefore dominate in a mixed population of recombinants. However, the most replicable molecule may not necessarily represent the most frequently formed recombinant. Consequently, multiple independent events and careful controls should be performed to confirm putative deletion hot-spots. Ideally, these events should be studied in systems which do not allow for appreciable amplification of newly formed molecules. The in vitro system developed by SONG and SIMON (1994, 1995), which uses an extract containing TCV-dependent polmerase activity, should allow for more definitive mapping of true recombination hot-spots and permit more detailed studies on the role of the template and polymerase in these processes.

3.4 Evolution of Structures

Differently sized DI RNAs have been observed in TBSV infections, and the examination of various structures has led to the suggestion that smaller molecules may be formed from larger ones (BURGYAN et al. 1991; KNORR et al. 1991; KNORR and MORRIS 1991). This concept appears to be correct, as various studies have shown that larger DI RNAs can evolve to smaller forms during serial passage (WHITE and MORRIS 1994b, 1995; HAVELDA et al. 1997). A stepwise deletion model has been proposed to describe the order of deletion events which lead to the formation of a prototypical *Tombusvirus* DI RNA (WHITE and MORRIS 1994b). In addition to standard large deletions, other types of modification, including segment duplication and nucleotide insertion, were identified during serial passage of precursor TBSV DI RNAs (WHITE and MORRIS 1994a). This observation indicates that these molecules not only can evolve to smaller forms via deletion but may also become larger by duplicating certain regions. DI RNAs may undergo significant modification after their initial formation; however, only those alterations which provide a selective advantage will be maintained. By analyzing these differences in structure, one may be able to estimate the relative 'evolutionary age' of these molecules.

4 Replication

The ability of DI RNAs to be replicated indicates that their structure must possess promoter signals that are utilized by the viral replicase. An additional requirement for their amplification is that the viral polymerase be able to act in *trans* on the DI template. That is, the replicase must be capable of amplifying a replicon other than the viral genome from which it was translated. A further requirement is that amplification of these molecules must be relatively efficient, so as to allow for their successful accumulation in the presence of the wild-type genome. In fact, many DI RNAs replicate more efficiently than their parental genome, and this situation is thought to contribute to observed decreases in accumulation levels of the parental genome.

4.1 Mechanisms of Replication

The genomes of the parental viruses discussed in this review are linear single-stranded (ss) RNAs. Their associated DI RNAs are also linear ssRNAs which are believed to replicate by a mechanism similar to that proposed for their parental counterparts; i.e., a negative strand is synthesized from the positive DI RNA, which in turn acts as a template for the synthesis of more positive strands. As with replication of the full-length genome, this process is predicted to be asymmetrical, with more plus strands being synthesized than minus strands. Evidence in support of asymmetry in DI RNA replication has been found by analyzing the accumulation levels of TBSV DI RNA plus and minus strands, where plus-strand accumulation was found to be several-fold higher (Ray and White, unpublished data). The majority of DI RNAs contain terminal regions of the genome which likely harbor important replication elements. However, in some cases internal sequences from the genome are also retained, and may play important roles in replication. Interestingly, head-to-tail dimer forms of two *Tombusvirus* DI RNAs have also been identified and characterized (Finnen and Rochon 1995; Dalmay et al. 1995). It has been suggested that these plus-sense dimers could represent important intermediates in the replication of monomers; however, an essential role for these molecules in the replication process has not yet been demonstrated.

4.2 Promoters

Various segments present in D and DI RNA molecules are highly conserved and must contain promoter sequences important for amplification. The maintenance of promoter elements makes these molecules particularly well suited for studies defining sequences and structures important for replication. It is, however, unclear as to whether the sequences conserved in these molecules match precisely those that are responsible for replication of the viral genome. The sequences in D and DI RNAs may represent truncated or incomplete copies of promoters which may be lacking important regulatory elements. Furthermore, the segments are present in different contexts and/or may be structurally modified from their original form, both of which could influence activity. Despite these possible differences, it is likely that replication elements in D and DI RNAs share significant functional features with those of the viral genome.

D RNAs of the potexviruses and closteroviruses contain only 5' and 3' termini of the genome, suggesting that essential promoter elements reside primarily in these terminal regions (MAWASSI et al. 1995a,b; WHITE et al. 1991; CALVERT et al. 1996). At present, it is unclear whether these conserved regions represent the minimal promoter elements, since further delineation of essential regions is yet to be performed. In the case of the *Potexvirus* D RNAs, which require an encoded fusion ORF for viability, it is possible that promoter activity is somehow coupled to translation of the ORF. Alternatively, translation of the fusion ORFs may lead to stabilization of the D RNAs. Only one putative promoter element, termed the hexanucleotide motif, has been identified in CYMV D RNAs. No accumulation of D RNAs was observed when this motif was disrupted (WHITE et al. 1992b), consistent with its possible role as a promoter element.

In *Tombusvirus* DI RNAs, conserved segments derived from terminal regions of the genome have been shown to be essential for viability and likely represent promoter elements (HAVELDA et al. 1995; HAVELDA and BURGYAN 1995; CHANG et al. 1995). Interestingly, conserved segments corresponding to internal regions of the viral genome may also be important for replication, since their deletion either prevents or greatly reduces levels of DI RNA accumulation (HAVELDA et al. 1995; CHANG et al. 1995). Studies on CyRSV DI RNAs revealed that structures in the 3'-terminal region of the DI RNA which are essential for DI RNA viability are also important in genome accumulation (HAVELDA and BURGYAN 1995). These studies confirmed that functionally important sequences and structures in the DI RNAs are also important in the virus genome.

In addition to sequence, size of the DI RNA has been shown to be important in the viability of DIs associated with TCV infections (LI and SIMON 1993). Replacement of deleted regions in a nonfunctional DI RNA with nonviral sequence led to restored viability, suggesting a spacing or size requirement.

The presence of promoter regions in DI RNAs has made them useful molecules for studying viral promoter elements. Although analysis of these molecules has allowed for the identification of sequences likely important for genome replication, most of these studies have been carried out using in vivo systems which do not allow for differentiation between effects on replication versus RNA stability.

5 Interference

D RNA molecules do not appreciably alter symptoms or virus yield, but their counterparts, DI RNAs, are able to interfere with various aspects of the infection process. The actual mechanism(s) by which these molecules are able to exert control over a virus infection is not well understood. One logical explanation for observed decreases in virus accumulation is that the DI RNA molecules are able to compete with the genome for limiting factors such as replicase (JONES et al. 1990). The overall decrease in virus may in turn lead to a reduction in the severity of symptoms. This proposal cannot, however, adequately explain all of the effects caused by DI RNAs.

5.1 Virus and Viral RNA Accumulation

D RNAs of ClYMV replicate efficiently and are encapsidated by the virus coat protein, yet no appreciable decrease in genome or sgRNA accumulation is observed when they are present in an infection (WHITE et al. 1991). Analysis of the yield of virus particles suggested that the amount of virus produced decreases when D RNAs were present in the infection, but the symptoms of D RNA-containing and D RNA-free infections were indistinguishable (WHITE et al. 1991). Similarly, no obvious differences in symptoms were detected in infections with or without CTV D RNAs (MAWASSI et al. 1995b). Currently, it is unclear as to why certain classes of defective molecules do not lead to a noticeable change in symptomology.

Conversely, studies on *Tombusvirus* DI RNAs have demonstrated that when they are present in an infection, there is a substantial decrease in the yield of authentic viral RNAs (HILLMAN et al. 1987; BURGYAN et al. 1989; ROCHON 1991). This observation was first made in infections of whole plants but was later confirmed in protoplast infections where a clear dose response was seen between an increase in DI RNA concentration in the inoculum and a corresponding decrease in the accumulation of viral RNA (JONES et al. 1990). It was suggested that this effect on authentic viral RNAs was due to efficient competition by the DI RNAs for a limited supply of replicase (JONES et al. 1990). Additional studies, went on to show that the suppression of authentic viral RNAs tended to be specific, in that sgRNAs seemed to be preferentially targeted for inhibition (SCHOLTHOF et al. 1995b). Similar decreases in authentic viral RNA accumulation were reported for coinfections of TCV and its DI RNAs, but no targeted inhibition of sgRNAs was observed (LI et al. 1989).

5.2 Symptomology

The virus-host interactions which lead to symptom induction are poorly understood. It therefore follows that a clear understanding of how DI RNAs influence symptomology is also lacking. While *Tombusvirus* infections cause lethal necrosis in various hosts, the presence of DI RNAs in such infections can lead to persistent attenuated (nonlethal) infections (RUSSO et al. 1994). For TBSV, it is known that both the p19 and p22 viral products are responsible for induction of symptoms (SCHOLTHOF et al. 1995a); but their mode of action has not yet been determined. In subsequent studies, it was also shown that TBSV DI RNAs caused reduced levels of sgRNAs encoding the capsid and nested genes in protoplasts and their corresponding encoded proteins. In contrast, a much less dramatic reduction was noted in the replicase (p33 and p92) proteins and in the level of genomic RNA. These results suggest that the protective effects exerted may be due in part to selective inhibition of p19 and p22 expression and in part to reduced replication of genomic RNA. The reduced levels of p19 could in turn result in much less extensive necrosis and a higher proportion of surviving plants (SCHOLTHOF et al. 1995b).

Interestingly, for TCV the presence of the DI RNAs, which reduce virus yields, results in increased rather than decreased symptom severity (LI et al. 1989). It was hypothesized that a segment within the DI RNA represents a symptom determinant and that the multiple copies produced by DI RNA amplification lead to an increased dose and more severe symptoms. More recently, the symptom response caused by TCV DI RNA G has been linked to the presence of the TCV coat protein (KONG et al. 1997). It was suggested that the coat protein may interact with the DI RNA and influence its symptom-modulating ability.

6 Applications

DI RNAs have several attributes which make them attractive candidates for various practical applications. Most of these molecules are amplified to very high levels, a feature which is favorable for use as expression vectors. It may therefore be possible to genetically engineer these molecules to encode and express a foreign protein of interest in plants. A second potentially useful feature of these molecules is their symptom-attenuating ability. Efforts have been made to take advantage of this attribute by producing DI RNA-expressing transgenic plants which would be protected from the lethal effects of an infection.

6.1 Expression Vectors

Studies with the full-length TBSV genome have shown that it can be engineered to express foreign proteins (SCHOLTHOF et al. 1993). Early studies with DI RNAs of CyRSV indicated that segments of nonviral sequence could be stably inserted into certain regions of these molecules (BURGYAN et al. 1992). Further tests showed that modified CyRSV DI RNA could be engineered to express the coat protein of an unrelated virus (BURGYAN et al. 1994). Although the extreme replicability of these molecules makes them attractive for expression, their unstable nature may ultimately limit their usefulness. With no positive selection pressure to maintain a foreign sequence, mutations which inactivate or remove the sequence will likely be maintained or even favored in a mixed population. However, this system may be useful for transient high-level expression in initially infected tissue.

6.2 Plant Protection

Attempts have been made to take advantage of the symptom-attenuating properties of DI RNAs for plant protection. Plants which transgenically express a CyRSV DI RNA showed reduced severity of symptoms upon challenge with virus; however, complete protection was not observed (KOLLAR et al. 1993). A drawback to this approach is the relatively high level of specificity, whereby partial protection would be observed only if the infecting virus were closely related. In contrast, a broader spectrum of resistance to several tombusviruses has been observed in transgenic plants expressing TBSV DI clones (RUBIO et al. 1998). This strategy incorporated ribozymes from satellite tobacco ringspot virus and avocado sunblotch viroid to flank the DI sequence at the 3′ and 5′ termini, so that wild-type DI RNA transcripts would be released in the transgenic plants. The transgenic plants exhibited a high level of resistance to the lethal necrosis caused by wild-type virus. Although a reduction in symptoms was noted in both studies, the challenged transgenic plants remain infected and could thus serve as reservoirs for further spread of the infection. This could limit the usefulness of such strategies for field applications.

7 Perspectives

Although genetic recombination in RNA viruses serves various useful functions such as repair and adaptation, the biological roles of D and DI RNAs are less clear. These molecules likely represent products of erroneous replication and serve no useful purpose. Nonetheless, they have been and will continue to be very useful molecules for studying important viral processes such as replication and recombination. Furthermore, once a more thorough understanding of the molecular biology of these molecules is achieved, the feasibility of developing commercially viable applications, such as foreign protein production and disease resistance, should increase significantly.

Acknowledgement. We thank Laurie Baggio for reviewing the manuscript.

References

Adam G, Gaedigk K, Mundry KW (1983) Alterations of a plant rhabdovirus during successive mechanical transfers. Z Pflanzenkr Pflanzensch 90:28–35
Burgyan J, Grieco F, Russo M (1989) A defective interfering RNA molecule in cymbidium ringspot virus infections. J Gen Virol 70:235–239
Burgyan J, Rubino L, Russo M (1991) De novo generation of cymbidium ringspot virus defective interfering RNA. J Gen Virol 72:505–509

Burgyan J, Dalmay T, Rubino L, Russo M (1992) The replication of cymbidium ringspot tombusvirus defective interfering-satellite RNA hybrid molecules. Virology 190:579 586

Burgyan J, Salanki K, Dalmay T, Russo M (1994) Expression of homologous and heterologous viral coat protein-encoding genes using recombinant DI RNA from cymbidium ringspot tombusvirus. Gene 138:159-163

Calvert LA, Cuervo MI, Ospina MD, Fauquet CM, Ramirez B (1996) Characterization of cassava common mosaic virus and a defective RNA species. J Gen Virol 77:525 530

Carpenter CD, Oh J, Zhang C, Simon AE (1995) Involvement of a stem-loop structure in the location of junction sites in viral RNA recombination. J Mol Biol 245:608-622

Cascone PJ, Carpenter CD, Li XH, Simon AE (1990) Recombination between satellite RNAs of turnip crinkle virus. EMBO J 9:1709-1715

Cascone PJ, Haydar TF, Simon AE (1993) Sequences and structures required for recombination between virus-associated RNAs. Science 260:801-805

Celix A, Rodriguez-Cerezo E, Garcia-Arenal F (1997) New satellite RNAs, but not DI RNAs, are found in natural populations of tomato bushy stunt virus. Virology 239:277-284

Chang YC, Borja M, Scholthof HB, Jackson AO, Morris TJ (1995) Host effects and sequences essential for accumulation of defective interfering RNAs of cucumber necrosis and tomato bushy stunt tombusviruses. Virology 210:41 53

Chetverin AB, Chetverina HV, Demidenko AA, Ugarov VI (1997) Nonhomologous RNA recombination in a cell-free system: evidence for a transesterification mechanism guided by secondary structure. Cell 88:503 513

Dalmay T, Szittya G, Burgyan J (1995) Generation of defective interfering RNA dimers of cymbidium ringspot tombusvirus. Virology 207:510–517

Dolja VV, Darasev AV, Koonin EV (1994) Molecular biology and evolution of closteroviruses: sophisticated build-up of large RNA genomes. Annu Rev Phytopathol 32:261 285

Figlerowicz M, Nagy PD, Bujarski JJ (1997) A mutation in the putative RNA polymerase gene inhibits nonhomologous, but not homologous, genetic recombination in an RNA virus. Proc Natl Acad Sci USA 94:2073 2078

Finnen RL, Rochon DM (1993) Sequence and structure of defective interfering RNAs associated with cucumber necrosis virus infections. J Gen Virol 74:1715–1720

Finnen RL, Rochon DM (1995) Characterization and biological activity of DI RNA dimers formed during cucumber necrosis virus coinfections. Virology 207:282–286

Graves MV, Pogany J, Romero J (1996) Defective interfering RNAs and defective viruses associated with multipartite RNA viruses of plants. Semin Virol 7:399–408

Havelda Z, Burgyan J (1995) 3' Terminal putative stem-loop structure required for the accumulation of cymbidium ringspot viral RNA. Virology 214:269–272

Havelda Z, Dalmay T, Burgyan J (1995) Localization of cis-acting sequences essential for cymbidium ringspot tombusvirus defective interfering RNA replication. J Gen Virol 76:2311 2316

Havelda Z, Dalmay T, Burgyan J (1997) Secondary structure-dependent evolution of cymbidium ringspot virus defective interfering RNA. J Gen Virol 78:1227–1234

Hillman BI, Schlegel DE, Morris TJ (1985) Effects of low molecular weight RNA and temperature on tomato bushy stunt virus symptom expression. Phytopathology 75:361 365

Hillman BI, Carrington JC, Morris TJ (1987) A defective interfering RNA that contains a mosaic of a plant virus genome. Cell 51:427–433

Inoue AK, Kormelink R, Nagata T, Kitajima EW, Goldbach R, Peters D (1997) Temperature and host effects on the generation of tomato spotted wilt virus defective interfering RNAs. Phytopathology 87:1168–1173

Jones RW, Jackson AO, Morris TJ (1990) Defective-interfering RNAs and elevated temperatures inhibit replication of tomato bushy stunt virus in inoculated protoplasts. Virology 176:539 545

Knorr DA, Morris TJ (1991) Origin and evolution of defective interfering RNAs of tomato bushy stunt virus. In: Herrmann RG, Larkins B (eds) Plant molecular biology, 2. Plenum, New York, pp 57-66

Knorr DA, Mullin RH, Hearne PQ, Morris TJ (1991) De novo generation of defective interfering RNAs of tomato bushy stunt virus by high multiplicity passage. Virology 181:193-202

Kollar A, Dalmay T, Burgyan J (1993) Defective interfering RNA-mediated resistance against cymbidium ringspot tombusvirus in transgenic plants. Virology 193:313 318

Kong Q, Oh J, Carpenter CD, Simon AE (1997) The coat protein of turnip crinkle virus is involved in subviral RNA-mediated symptom modulation and accumulation. Virology 238:478 485

Koonin EV, Dolja VV (1993) Evolution and taxonomy of positive-strand RNA viruses: implications of comparative analysis of amino acid sequences. CRC Crit Rev Biochem Mol Biol 28:375 430

Lai MMC (1992) RNA recombination in animal and plant viruses. Microbiol Rev 56:61-79

Law MD, Morris TJ (1994) De novo generation and accumulation of tomato bushy stunt virus defective interfering RNAs without serial host passage. Virology 198:377-380

Lazzarini RA, Keene JD, Schubert M (1981) The origins of defective interfering particles of the negative-strand RNA viruses. Cell 26:145-154

Li XH, Simon AE (1993) In vivo accumulation of a turnip crinkle virus defective interfering RNA is affected by alterations in size and sequence. J Virol 65:4582-4590

Li XH, Heaton LA, Morris TJ, Simon AE (1989) Turnip crinkle virus defective interfering RNAs intensify viral symptoms and are generated de novo. Proc Natl Acad Sci USA 86:9173-9177

Mawassi M, Darasev AV, Mietkiewska E, Gafny R, Lee RF, Dawson WO, Bar-Joseph M (1995a) Defective RNA molecules associated with citrus tristeza virus. Virology 208:383-387

Mawassi M, Mietkiewska E, Hilf ME, Ashoulin L, Karasev AV, Gafny R, Lee RF, Garnsey SM, Dawson WO, Bar-Joseph M (1995b) Multiple species of defective RNAs in plants infected with citrus tristeza virus. Virology 214:264-268

Nagy PD, Bujarski JJ (1993) Targeting the site of RNA-RNA recombination in brome mosaic virus with antisense sequences. Proc Natl Acad Sci USA 90:6390-6394

Nagy PD, Bujarski JJ (1995) Efficient system of homologous RNA recombination in brome mosaic virus: sequences and structure requirements and accuracy of crossovers. J Virol 69:131-140

Nagy PD, Bujarski JJ (1997) Engineering of homologous recombination hotspots with AU-rich sequences in brome mosaic virus. J Virol 71:3799-3810

Nagy PD, Simon AE (1997) New insights into the mechanisms of RNA recombination. Virology 235:1-9

Nagy PD, Carpenter CD, Simon AE (1997) A novel 3'-end repair mechanism in an RNA virus. Proc Natl Acad Sci USA 94:1113-1118

Perrault J (1981) Origin and replication of defective interfering particles. In: Compans RW, Cooper M, Koprowski H, et al (eds) Current topics in microbiology and immunology, vol 93. Springer, Berlin Heidelberg New York, pp 151-207

Rochon DM (1991) Rapid de novo generation of defective interfering RNA by cucumber necrosis virus mutants that do not express the 20-kDa nonstructural protein. Proc Natl Acad Sci USA 88:11153-11157

Romero J, Huang Q, Pogany J, Bujarski JJ (1993) Characterization of defective interfering RNA components that increase symptom severity of broad bean mottle virus infections. Virology 194:576-584

Roossinck MJ, Sleat D, Palukaitis P (1992) Satellite RNAs of plant viruses: structures and biological effects. Microbiol Rev 56:265-279

Rubino L, Russo M (1997) Molecular analysis of the pothos latent virus genome. J Gen Virol 78:1219-1226

Rubino L, Burgyan J, Russo M (1995) Molecular cloning and complete nucleotide sequence of carnation Italian ringspot tombusvirus genomic and defective interfering RNAs. Arch Virol 140:2027-2039

Rubio T, Borja M, Scholthof HB, Feldstein P, Bruening G, Morris TJ, Jackson AO (1998) Broad spectrum resistance to tombusviruses elicited by defective interfering RNAs in transgenic plants (in preparation)

Russo M, Burgyan J, Martelli GP (1994) Molecular biology of Tombusviridae. Adv Virus Res 44:381-428

Scholthof HB, Morris TJ, Jackson AO (1993) The capsid protein gene of tomato bushy stunt virus is dispensable for systemic movement and can be replaced for localized expression of foreign genes. Mol Plant Microbe Interact 6:309-322

Scholthof HB, Scholthof KB, Jackson AO (1995a) Identification of tomato bushy stunt virus host-specific symptom determinants by expression of individual genes from a potato virus X vector. Plant Cell 7:1157-1172

Scholthof KB, Scholthof HB, Jackson AO (1995b) The effect of defective interfering RNAs on the accumulation of tomato bushy stunt virus proteins and implications for disease attenuation. Virology 211:324-328

Simon AE, Howell SH (1986) The virulent satellite RNA of turnip crinkle virus has a major domain homologous to the 3' end of the helper virus genome. EMBO J 5:3423-3428

Song C, Simon AE (1994) RNA-dependent RNA polymerase from plants infected with turnip crinkle virus can transcribe (+)- and (−)-strands of virus-associated RNAs. Proc Natl Acad Sci USA 91:8792-8796

Song C, Simon AE (1995) Synthesis of novel products in vitro by an RNA-dependent RNA polymerase. J Virol 69:4020-4028

Verkleij F, Peters D (1983) Characterization of a defective form of tomato spotted wilt virus. J Gen Virol 64:677–686

White KA (1996) Formation and evolution of tombusvirus defective interfering RNAs. Semin Virol 7:409–416

White KA, Morris TJ (1994a) Enhanced competitiveness of tomato bushy stunt virus defective interfering RNAs by segment duplication or nucleotide insertion. J Virol 68:6092–6096

White KA, Morris TJ (1994b) Nonhomologous RNA recombination in tombusviruses: generation and evolution of defective interfering RNAs by stepwise deletions. J Virol 68:14–24

White KA, Morris TJ (1995) RNA determinants of junction site selection in RNA virus recombinants and defective interfering RNAs. RNA 1:1029–1040

White KA, Bancroft JB, Mackie GA (1991) Defective RNAs of clover yellow mosaic virus encode nonstructural/coat protein fusion products. Virology 183:479–486

White KA, Bancroft JB, Mackie GA (1992a) Coding capacity determines in vivo accumulation of a defective RNA of clover yellow mosaic virus. J Virol 66:3069–3076

White KA, Bancroft JB, Mackie GA (1992b) Mutagenesis of a hexanucleotide sequence conserved in potexvirus RNAs. Virology 189:817–820

Yang G, Mawassi M, Ashoulin L, Gafny R, Gava V, Gal-On A, Bar-Joseph M (1997a) A cDNA clone from a defective RNA of citrus tristeza virus is infective in the presence of the helper virus. J Gen Virol 78:1765–1769

Yang G, Mawassi M, Gofman R, Gafny R, Bar-Joseph M (1997b) Involvement of a subgenomic mRNA in the generation of a variable population of defective citrus tristeza virus molecules. J Virol 71:9800–9802

Replication, Recombination, and Symptom-Modulation Properties of the Satellite RNAs of Turnip Crinkle Virus

A.E. SIMON

1 Introduction

Turnip crinkle virus (TCV) is a member of the carmovirus group and the only group member that is associated with confirmed sat-RNAs. The host range of TCV is quite widespread, including numerous cruciferous and some noncruciferous hosts (BROADBENT and HEATHCOTE 1958). TCV has a single RNA genome of 4054 bases (CARRINGTON et al. 1989; OH et al. 1995) and two subgenomic RNAs of 1721 bases and 1447 bases (CARRINGTON et al. 1987; WANG and SIMON 1997) (Fig. 1). The genomic RNA is the mRNA for two overlapping polypeptides required for viral replication, p28 and p88. The translation of p88, which contains the conserved polymerase active site motif GDD, requires an in-frame ribosomal readthrough of an amber codon at the end of the p28 open reading frame (ORF). Two small ORFs (p8 and p9), have been implicated in viral movement (HACKER et al. 1992) and are

Department of Biochemistry and Molecular Biology, University of Massachusetts, Amherst, MA 01003, USA

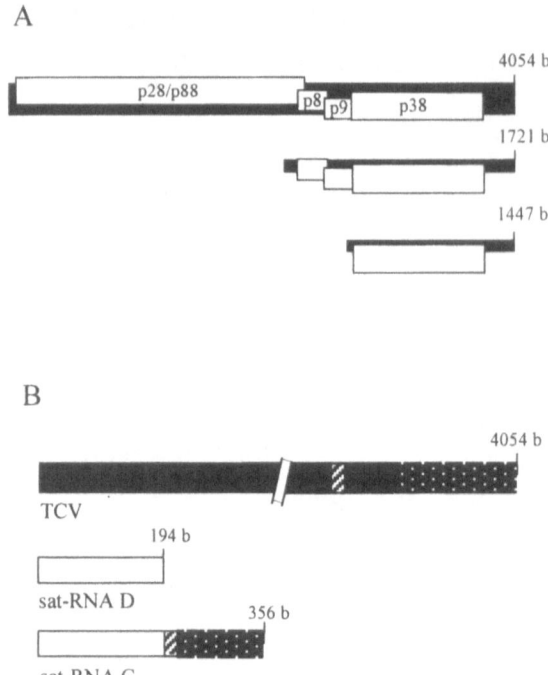

Fig. 1A,B. The TCV system. **A** The genomic and subgenomic RNAs of TCV. **B** Sequence relationships among the TCV genomic RNA and two sat-RNAs. Similar sequences are *shaded* alike. All TCV-associated RNAs also share a seven-base motif at the 3′ end, CCUGCCC-3′

translated in vitro from the 1.7-kb subgenomic RNA (T.J. Morris, personal communication). The virus coat protein (CP) p38 is translated from the 1.45-kb subgenomic RNA. The TCV virion is icosahedral in structure with T = 3 symmetry and is composed of 180 copies of the 38-kDa CP (HOGLE et al. 1986; CARRINGTON et al. 1989). The encapsidation signal on the genomic RNA is contained within a 186-base region at the 3′ end of the CP ORF and includes an essential 28-base hairpin (QU and MORRIS 1997). The lack of detectable packaging of wild-type subgenomic RNAs or of a genomic RNA that was 600 bases larger than wild-type suggests that the size of the RNA is an important factor for packaging into the icosahedral virion (QU and MORRIS 1997). Since this packaging sequence is not present in two highly infectious subviral RNAs, sat-RNA D and DI RNA G, other sequences must also be able to support a packaging function.

TCV is associated with numerous subviral RNAs (Fig. 1). The TCV-M isolate contains three subviral RNAs that were identified as sat-RNAs since they required the helper virus for amplification in plants and contained substantial regions without sequence similarity to the genomic RNA (ALTENBACH and HOWELL 1981; SIMON and HOWELL 1986; SIMON et al. 1989). Sat-RNA D is a typical sat-RNA, with little obvious sequence similarity with the TCV genomic RNA beyond the seven-base 3′-terminal motif (CCUGCCC-3′) that is found at the 3′ termini of all

RNAs associated with TCV (Simon and Howell 1986). Sat-RNA F (230 bases) is closely related to sat-RNA D, with the exception of a 36-base insert located near the 3′ end of sat-RNA F. Sat-RNA C, the most studied of the three sat-RNAs, is a hybrid molecule composed of a sat-RNA D-related sequence at the 5′ end (88% similarity) and two regions of TCV at the 3′ end (90% sequence similarity). The TCV-B isolate is associated with an RNA the size of sat-RNA D that hybridizes to sat-RNA D probes and with a defective interfering (DI) RNA (DI RNA G) that is composed mainly of helper virus sequence and the 5′-terminal ten bases of the TCV-M sat-RNAs (Li et al. 1989).

2 Symptom Modulation by TCV Sat-RNAs

A number of sat-RNAs associated with plant viruses are able to modify the symptoms expressed by their helper virus. Sat-RNA D and sat-RNA F have little or no effect on virus symptoms (Simon and Howell 1986; Simon et al. 1989; Li and Simon 1990). Sat-RNA C, however, can either intensify or attenuate symptoms depending on the host and the level of CP produced by the TCV genomic RNA (Simon et al. 1988; Li and Simon 1990; Kong et al. 1995; J. Wang and A.E. Simon, unpublished). In hosts that are tolerant to TCV infection (and therefore have no symptoms associated with virus infection), the presence of sat-RNA C has no effect on plant symptomology (Li and Simon 1990). In plants that normally develop symptoms in response to infection with the TCV genomic RNA the symptoms are intensified by the addition of sat-RNA C. For example, the normal slight stunting and slightly mottled and crinkled leaves associated with TCV infection of the turnip is intensified in the presence of sat-RNA C to a severe stunting with dark green, very crinkled leaves (Simon et al. 1988; Li et al. 1989). In most ecotypes of *Arabidopsis thaliana*, TCV causes moderate stunting, while the presence of the sat-RNA results in the death of the plant by about 16 days postinoculation (Li and Simon 1990; Simon et al. 1992). Whole plant in situ hybridizations have shown that the TCV genomic RNA becomes concentrated in younger tissue in the presence of the sat-RNA, leading to an inhibition of bolting and plant death (Q. Kong and A.E. Simon, unpublished).

    The involvement of the TCV CP in sat-RNA C symptoms has been demonstrated using several TCV derivatives. TCV that contains the CP of the related virus cardamine chlorotic fleck in place of the TCV CP (TCV-CP$_{CCFV}$) is infectious on the susceptible *Arabidopsis* ecotype Col-0, producing stunted plants (Oh et al. 1995). In the presence of sat-RNA C, the symptoms associated with TCV-CP$_{CCFV}$ disappeared, indicating that the normally virulent sat-RNA C can also attenuate viral symptoms (Kong et al. 1995). Northern analysis of RNA extracted from inoculated leaves at 7 or 10 days post inoculation revealed an 80% reduction in TCV-CP$_{CCFV}$ levels when co-inoculated with sat-RNA C with no detectable virus in uninoculated leaves (Kong et al. 1995, 1997b). In *Arabidopsis* protoplasts at 24 h postinoculation, the presence of sat-RNA C resulted in a 70% reduction in the level of TCV-CP$_{CCFV}$ (KONG et al. 1995), suggesting that sat-RNA C was competing with TCV-CP$_{CCFV}$ for factors required for genomic RNA amplification.

Sat-RNA C is also able to attenuate the symptoms of a TCV variant that contained a single base mutation in the initiation codon of the CP ORF (AUG to ACG) (TCV-CPm). Surprisingly, TCV-CPm produced systemic symptoms (stunted, sometimes bushy plants with curled cauline leaves) on Col-0, even though TCV requires the CP for movement (HACKER et al. 1992; LAAKSO and HEATON 1993) and the next in-frame AUG is 40 amino acids downstream from the initiation codon. Plants and protoplasts infected with TCV-CPm accumulate only 20% of the wild-type level of a protein that migrates slightly slower than wild-type TCV CP in polyacrylamide gels and cross-reacts with TCV CP-specific antibodies (KONG et al. 1997b). Attenuation of TCV-CPm symptoms by sat-RNA C is correlated with little or no detectable genomic RNA in uninoculated leaves. Unlike TCV-CP$_{CCFV}$, sat-RNA C did not substantially affect the accumulation of TCV-CPm genomic RNA in protoplasts. Since TCV-CPm was present in inoculated leaves in the presence or absence of sat-RNA C, sat-RNA C likely affects the movement of TCV into uninoculated leaves (KONG et al. 1997b).

As TCV-CPm differs from TCV by only a single base alteration, the CP and not the RNA encoding the CP was thought to be the viral determinant that affects the symptoms of sat-RNA C. Since virions were nearly undetectable in preparations of TCV-CPm-infected protoplasts (KONG et al. 1997b), it is probable that the TCV-CPm CP is not wild type and contains an altered N-terminus due to initiation at a non-AUG codon in the vicinity of the natural initiation codon. The ability of TCV-CPm to promote symptom attenuation by sat-RNA C could therefore be due either to the putative mutation in the amino terminus of the TCV-CPm CP or to the presence of reduced levels of CP present in cells infected with TCV-CPm. By altering the promoter for the 1.45-kb subgenomic RNA, TCV variants were produced that express levels of wild-type CP similar to TCV-CPm (~10–20% of wild type). Since these mutant TCV also have their symptoms attenuated by sat-RNA C (J. WANG and A.E. SIMON, unpublished), the level of CP is the crucial factor in deciding whether the sat-RNA is going to attenuate or intensify symptoms. The region of sat-RNA C involved in such symptom modulation has been localized to the 3′-terminal 53 bases (KONG et al. 1997a).

To determine if the salicylic acid-dependent defense pathway is involved in sat-RNA C-mediated resistance, several Col-0-derived lines with presumptive mutations in pathway enzymes were tested for symptom production by TCV, TCV-CP$_{CCFV}$, and TCV-CPm, in the presence and absence of sat-RNA C. These mutant plants behaved like wild-type Col-0 plants with respect to symptoms, and by analogy, to systemic movement of TCV, TCV-CP$_{CCFV}$, and TCV-CPm in the presence of sat-RNA C (KONG et al. 1997b). In addition, there was no pathogenesis-related (PR) protein mRNA accumulation above basal levels in inoculated leaves of Col-0 infected with TCV-CPm and sat-RNA C and no apparent induction of systemic acquired resistance (KONG et al. 1997b). These results indicate that the major defense pathway in *Arabidopsis* is not involved in symptom attenuation by sat-RNA C.

The TCV RNA-dependent RNA polymerase (RdRp) can also affect the symptoms mediated by sat-RNA C. Alterations at several positions in p88 decreased the intensity of symptoms produced by the virus in the presence of sat-RNA C (COLLMER et al. 1992; OH et al. 1995); this may have been due to a decrease in the early rate of sat-RNA C accumulation in infected cells (COLLMER et al. 1992).

3 Replication of TCV Sat-RNAs

Replication of the genome of a plus-strand. RNA virus proceeds through a complementary minus-strand intermediate and is catalyzed by the virus-encoded RdRp. Circular sat-RNAs or sat-RNAs that have a circular form inside cells (e.g., satellite of tobacco ringspot virus) produce multimeric minus strands by a rolling circle mechanism, followed by RNA-mediated cleavage to the final monomeric form (PRODY et al. 1986; FORSTER and SYMONS 1987). TCV sat-RNA C and sat-RNA D, which do not have detectable circular forms in cells (A.E. Simon, unpublished), are still associated with substantial amounts of multimeric plus strands (ALTENBACH and HOWELL 1981; CARPENTER et al. 1991). Sequencing the junctions between monomeric units of sat-RNA C and sat-RNA D revealed that, unlike junctions found in rolling circle replication, the junction sequences were not a precise joining of the 3' and 5' ends but rather contained truncations and/or sequence duplications reminiscent of the junctions of recombinant RNAs (CARPENTER et al. 1991) (see Sect. 4). This result indicated that despite the accumulation of multimeric species, rolling circle replication does not take place and the probable mode of replication is through a linear unit length minus-strand intermediate.

3.1 *Cis*-Sequence Involved in Sat-RNA C Replication

Protoplast and in vitro systems have been developed to study the replication of TCV sat-RNAs. TCV RdRp activity has been solubilized from membranes and three peaks of activity can be separated following Sephacryl S500 HR chromatography (SONG and SIMON 1994). Peak I, the only peak that contains substantial amounts of endogenous TCV, and peak II are template specific, synthesizing full-length complementary strands of exogenous TCV sat-RNAs but not control RNA templates. Peak III is nonspecific, synthesizing full-sized products for all added RNA templates. Peak II RdRp accepts both plus- and minus-strand sat-RNA as templates, with the minus strand of a species being the more active template. Peak II RdRp has been used to define promoter sequences on the plus- and minus-strands of sat-RNA C. As described below, both hairpin structures and short primary sequences can serve as independent promoters in vitro.

3.1.1 *Cis*-Sequence Involved in the Synthesis of Sat-RNA C Minus Strands

Addition of full-length plus-strand sat-RNA C transcripts to the RdRp extract results in the synthesis of full-length minus strands and a second RNA that migrates slightly faster than full-length and is probably a product of internal initiation (SONG and SIMON 1995b). Using the in vitro system, the *cis*-sequences required for synthesis of minus-strand sat-RNA C from plus-strand template were delineated (SONG and SIMON 1995a). Relocating the 3'-terminal 37 bases to an inactive template (MDV-1 RNA associated with bacteriophage Qβ) rendered the resultant hybrid RNA competent for in vitro transcription by the RdRp, indicating that the promoter for minus-strand synthesis in vitro is completely contained within this 3'-terminal region. Additional deletion analysis showed that the promoter is contained

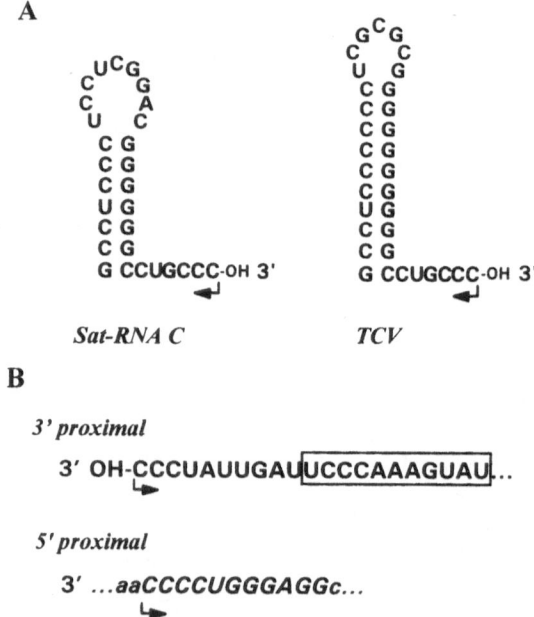

Fig. 2A,B. Sat-RNA C promoter sequences. **A** Hairpin promoter located at the 3' end of sat-RNA C plus strands. *Arrow* denotes the transcription initiation site. The similar sequence at the 3' end of TCV genomic RNA is also shown. **B** The two promoters located on sat-RNA C minus strands. *Top:* The 3'-proximal promoter is *boxed* and is the only 3' end sequence required for transcription in vitro. *Arrow* denotes the wild-type transcription initiation site. In the absence of the 3'-terminal six bases (CCCUAU), or if nontemplate nucleotides are added to the 3' end of the minus strand, transcription is initiated at the three C residues within the promoter sequence. *Bottom:* 5'-Proximal promoter on sat-RNA C minus strands. This promoter is located 41 bases from the 5' end of the minus strands. It is not known whether the nucleotides in *lowercase italics* are required for promoter function. *Arrow* denotes the major transcription initiation site in the absence of the 3'-proximal promoter

within the 3'-terminal 29 bases of the plus strand (SONG and SIMON 1995a). This region contains a stem-loop structure with a 3'-terminal six-base single-stranded tail, as revealed by both computer RNA secondary structure analysis and structural probing by enzymatic digestion and chemical modification (Fig. 2A). Mutagenesis of the stem and loop regions, followed by in vitro (SONG and SIMON 1995a) and in vivo (STUPINA and SIMON 1997) analyses, indicated that the primary sequence of the loop and stem are not essential for interaction with the viral RdRp. However, single mutations that disrupted the base of the stem or double/triple mutations in the upper stem substantially reduced template activity in vitro and accumulation in vivo, suggesting that the stability of the hairpin is important (SONG and SIMON 1995b). Monomeric-length sat-RNA C was synthesized from template containing as many as 220 nonsatellite bases at the 3' ends of plus strands, indicating that the RdRp can recognize the 3'-end promoter in an internal location (SONG and SIMON 1994).

The sequence/structure requirements of the sat-RNA C 3' end plus-strand promoter were also recently analyzed by a modification of the in vitro genetic selection procedure known as SELEX (Systematic Evolution of Ligands by EXponential enrichment) (ELLINGTON and SZOSTAK 1990; TUERK and GOLD 1990). This procedure involves the generation of a population of molecules containing randomized sequence, followed by multiple rounds of enrichment for molecules from the population that are more fit at performing a particular function such as binding to a specific protein or catalyzing an enzymatic reaction (GOLD et al. 1995). For analysis of the sat-RNA C promoter in vivo, the sequence in the 22-base hairpin portion of the promoter was randomized and the transcript population produced inoculated onto turnip plants with the TCV helper virus (CARPENTER and SIMON, 1998). Sat-RNA transcripts that contain viable promoters would be replicated by the viral RdRp and would then be present in uninoculated leaves. RNA extracted from uninoculated leaves of multiple plants were pooled and re-inoculated onto seedlings to initiate further "rounds" of direct competition between various promoter sequences. After three rounds of selection, four sequence family "winners" were revealed, with three families containing multiple variants, indicating that evolution of these sequences was occurring from the many rounds of replication required for sat-RNA C accumulation in uninoculated leaves (Fig. 3). Three of the four sequence family winners had the same three base pairs at the base of the stem as wild-type sat-RNA C. Two of the winners shared 15 of 22 identical bases, including the entire stem region and extending two bases into the loop. The sequence and structural similarities between these two winners and the sequence similarity at the base of the stem for nearly all third-round winners and wild-type sat-RNA C strongly suggest that both sequence and structure in the promoter region contribute to increased fitness of sat-RNA C.

3.1.2 Cis-Sequence Involved in the Synthesis of Sat-RNA C Plus Strands

One of the major advantages of the TCV system for the study of viral RNA replication is the ability to use minus-strand sat-RNAs as templates in in vitro

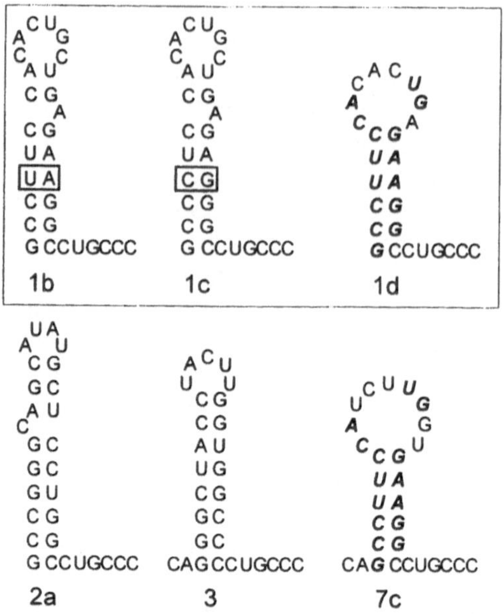

Fig. 3. "Winners" of the in vivo genetic selection for sequences that can form a functional sat-RNA C promoter. Three rounds of in vivo genetic selection were conducted, beginning with a population of molecules containing randomized hairpin sequences. Three variants of sequence family winner *1* are *boxed*. Differences between two of the variants are indicated. The *1b* sequence variant is favored in direct competition with the *1c* variant (CARPENTER and SIMON, 1998), suggesting that a weak base pair in the fourth position of the stem is favored (see the wild-type sat-RNA C hairpin in Fig. 2A). Sequence identity between two family winners (*1d* and *7c*) that arose independently is indicated by italics

transcription reactions catalyzed by the viral RdRp (SONG and SIMON 1994). Programming the in vitro extracts with minus-strand sat-RNA C results in the synthesis of three major products: (a) full-length plus strands; (b) a hairpin species composed of a full-length plus strand covalently linked to the minus-strand template (S-RNA); (c) a panhandle species formed from an internal primer extension reaction that results in a 190-base product covalently linked to the full-length minus-strand template (L-RNA) (SONG and SIMON 1994, 1995b). Synthesis of the L-RNA product requires the presence of a hairpin in sat-RNA C (motif1-hairpin) that is also required for RNA recombination (see Sect. 4.1) (SONG and SIMON 1995b).

Two sequences have been identified in sat-RNA C minus-strands that are able to function as independent promoters in vitro (GUAN et al. 1997). The 3'-proximal promoter sequence (Fig. 2B) is located 11 bases from the 3' end of the minus strands. The 3'-terminal ten nucleotides of sat-RNA C minus strands are not required for promoter activity; however, this region shares the consensus sequence at the 3' termini of minus strands of other carmovirus genomic and subgenomic RNAs [3' (C)CCA/U A/U A/U]. This same conserved sequence is also found at the 3' end of the 3'-proximal promoter of minus-strand sat-RNA C (3' CCCAAA).

Although the 3' terminal ten nucleotides of sat-RNA C minus strands are not required for promoter activity, these bases may be important in RdRp initiation of transcription at the 3' end of minus strands. Transcription in vitro is initiated internally at the 3'-proximal promoter's CCC residues if plasmid-derived bases are added to the 3' ends of minus-strand templates (GUAN et al. 1997). Deletion of positions 1–10 eliminates internal initiation in the presence of the plasmid-derived nucleotides.

A second sequence in sat-RNA C minus strands that can function as an independent promoter in vitro is located 41 bases from the 5' end (Fig. 2B) (GUAN et al. 1997). Eight bases of this sequence are conserved in the motif1-hairpin also located on minus-strand sat-RNA C, which is required for RNA recombination and may be an RdRp recognition site (CASCONE et al. 1990, 1993; P.D. NAGY, C. ZHANG and A.E. SIMON, manuscript submitted). The 5'-proximal promoter and 3'-proximal promoter sequences are short motifs without obvious secondary structure and contain multiple consecutive C residues followed by multiple consecutive purines. As with the 3'-proximal promoter, transcription can be initiated internally within the 5'-proximal promoter at the multiple C residues. Recent results suggest that the 5' proximal promoter participates in plus-strand synthesis in vivo (H. GUAN and A.E. SIMON, unpublished).

3.2 Role of the CP in Replication of TCV Sat-RNAs

In addition to the TCV RdRp, the viral-encoded CP also appears to have a role in the accumulation of TCV sat-RNAs. The level of sat-RNA C in infected protoplasts increased three- to fourfold when sat-RNA C was co-inoculated with virus that did not produce TCV CP (e.g., in the presence of TCV-CP$_{CCFV}$ or TCV with a deletion of the CP ORF) (KONG et al. 1995). This result suggests that the TCV CP represses the accumulation of sat-RNA C. Since the level of sat-RNA C accumulating in the presence of either wild-type TCV or TCV-CPm (which produces only 10–20% of the wild-type level of CP) was similar, less than wild-type levels of CP are needed to repress the accumulation of sat-RNA C (KONG et al. 1997b). Interestingly, the TCV CP has an opposite effect on the accumulation of sat-RNA D; the level of sat-RNA D accumulating in plants or protoplasts was more than ten times higher in the presence of wild-type TCV than with TCV-CP$_{CCFV}$ (KONG et al. 1997a). The specific role that the TCV CP plays in the replication mechanism has not been elucidated.

4 Recombination Among TCV Sat-RNAs

Recombination among viral RNAs has been documented for an increasing number of both plant and animal viruses (SIMON and BUJARSKI 1994). The majority of available evidence points to a replication-mediated template-switching mechanism for the generation of recombinants (NAGY and SIMON 1997). However, the molecular mechanisms that lead to the recovery of recombinant molecules have been studied in only a few viral systems due to the low frequency of crossover events and the scattered distribution of junction sites for many viruses. The uniquely high recombination frequency and nonrandom crossover site distribution for recombination between TCV sat-RNAs and between the sat-RNAs and the TCV genomic RNA has made TCV one of the best-studied recombination systems (SIMON and NAGY 1997). TCV is also the only eukaryotic viral system so far in which recombination has been analyzed using whole organism, cell culture, and in vitro approaches.

Junction sequences in TCV recombinant molecules frequently contain non-template bases that consist mainly of multiple uracil residues (CASCONE et al. 1990, 1993). The finding of similar junction sequences between monomeric units of multimeric sat-RNA species and at the junctions in molecules that have undergone 3'-end repair (see Sect. 5) suggest that a similar mechanism is involved in these events. Based on the sequence just downstream from the crossover site in numerous recombinant molecules including TCV DI RNAs (LI et al. 1989), the RdRp is thought to recombine RNAs mainly during the synthesis of plus strands (CASCONE et al. 1990; ZHANG et al. 1991).

4.1 Recombination Between Sat-RNA D and Sat-RNA C

Plants inoculated with TCV genomic RNA, sat-RNA D, and sat-RNA C that contain deletions in the 5' region accumulated a heterogeneous population of recombinant molecules comprising sat-RNA D at the 5' end and a portion of sat-RNA C at the 3' end (CASCONE et al. 1990, 1993). The junction between the sat-RNA D and sat-RNA C sequence occurred mainly at position 181 of sat-RNA D (13 bases from the 3' end of the plus strand) and one of five consecutive bases in sat-RNA C beginning at position 175 (CASCONE et al. 1990, 1993; Fig. 4A). Just downstream from the recombination site in sat-RNA C is a stable hairpin (motif1-hairpin) in minus strands of sat-RNA C (CARPENTER et al. 1995; Fig. 4). Mutations generated within the motif1-hairpin that disrupted the stem of the hairpin eliminated the detection of recombinants in plants, while compensatory mutations that restored the hairpin restored recombination (CASCONE et al. 1993). Alterations of some, but not all the bases in the loop also negatively affected the recovery of recombinants, as did deletion of the three U residues in the bulge (CASCONE et al. 1993; NAGY et al., in press). In addition, deletion of nucleotides located 16–20 bases upstream of the hairpin eliminated detectable recombinants in infected plants (CASCONE et al. 1993).

Recombination between sat-RNA D and sat-RNA C has also been studied using an in vitro recombination system (NAGY et al., 1998, Nagy et al., in press), which has provided direct evidence for a replicase-mediated template-switching mechanism for RNA recombination in TCV. The system is based on the model

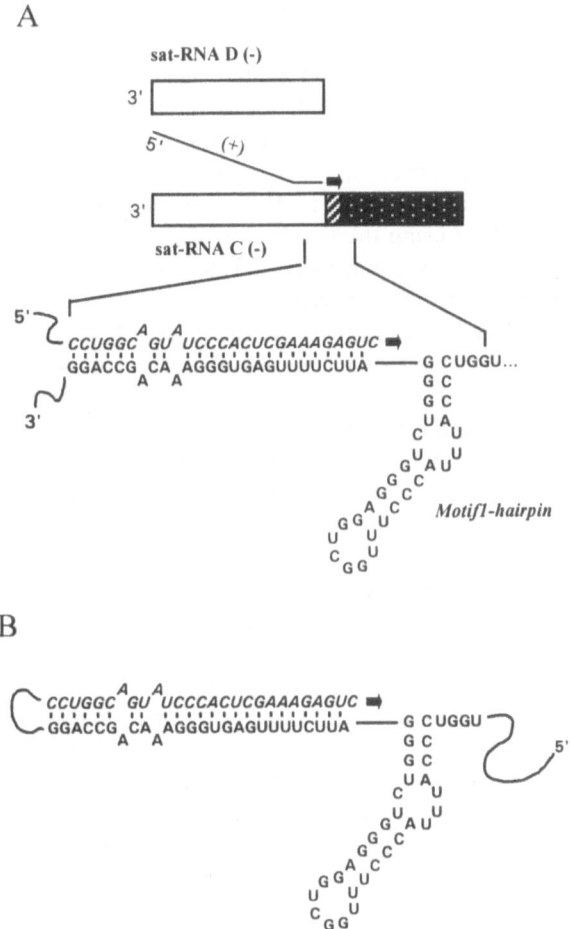

Fig. 4A,B. RNA recombination between sat-RNA D and sat-RNA C. **A** Model for recombination between sat-RNA D and sat-RNA C. After finishing or nearly finishing synthesis of sat-RNA D plus-strands (+), the RdRp switches to the minus-strand (−) sat-RNA C template and transcription continues, producing a recombinant molecule. The motif1-hairpin is thought to recruit the RdRp to the sat-RNA C minus-strand template. Base pairing between the nascent sat-RNA D plus strand and sat-RNA C minus strand positions the RdRp to reinitiate synthesis just 3′ of the motif1-hairpin. Sat-RNA D plus-strand sequence (truncated at the -13 "hot-spot" position) is shown in *italics*. **B** Template used for in vitro recombination studies. A short loop sequence joins the plus-strand sat-RNA D sequence to minus-strand sat-RNA C. "Recombination" is assayed by detection of primer-extension products covalently attached to the template (NAGY et al., 1998)

derived from the in vivo data, which suggests that the TCV RdRp, after synthesizing full-length or near full-length nascent sat-RNA D plus strands, switches to the sat-RNA C minus-strand template at the base of the motif1-hairpin and then transcribes the sat-RNA C plus strand covalently linked to the previously synthesized sat-RNA D plus strand (Fig. 4A). To reconstruct this system, a chimeric

RNA was generated containing (from 5′ to 3′): sat-RNA C minus strand from the 5′ end through the hairpin and some downstream sequences; a loop region; sat-RNA D plus-strand 3′ end sequence (Fig. 4B). This construct, which can form a nearly perfect duplex between the sat-RNA D plus-strand sequence and the sat-RNA C sequence downstream from the hairpin (the duplex region), allows for recombination without the need to bring the two templates together in solution. Recombination in vitro is detected as an RdRp-dependent extension product from the 3′ end of sat-RNA D using the sat-RNA C-derived region as template. Using this construct and other derived constructs, the need for the sat-RNA C hairpin, the importance of the duplex region between sat-RNA D and sat-RNA C, and the need for a spacer region between the duplex and the hairpin have been studied. Results indicate that (a) mutations in the hairpin that eliminate recombination in vivo also markedly decrease the synthesis of extension products in vitro; (b) the 3′-terminal region of the duplex plays a role in terminal extension; (c) mismatches in the duplex can affect both the amount of product and the site of terminal extension; (d) the length of the spacer between the hairpin and the duplex influences the efficiency of terminal elongation; (e) the motif1-hairpin is important in recruiting the RdRp to the 3′ end of the template. These results suggest that RNA recombination is similar to primer extension reactions by DNA-dependent RNA polymerases; however, extension of the "primer" requires a stable hairpin on the template strand, an aspect that may be unique to RNA-dependent RNA polymerases.

The finding that the motif1-hairpin may be involved in recruitment of the RdRp to the template prior to primer extension in vitro suggests that this structure may be able to serve as a general transcription enhancer for TCV RNAs. Recent results indicate that an approximate 30-fold increase in full-length synthesis of either sat-RNA D plus-strand or minus-strand transcripts can be obtained in vitro if the motif1-hairpin is incorporated into the sequence (P.D. NAGY and A.E. SIMON, unpublished)

4.2 Recombination Between TCV and Sat-RNA D

Plants inoculated with wild-type TCV and sat-RNA D accumulate a population of different recombinant molecules composed (from 5′ to 3′) of mainly full-length sat-RNA D or sat-RNA D truncated by 13 bases joined to variable lengths of the TCV 3′-end sequence (ZHANG et al. 1991; CARPENTER et al. 1995). To date, none of the sat-RNA D/TCV recombinants tested can be amplified to detectable levels in plants or protoplasts (CARPENTER et al. 1995; CARPENTER and SIMON 1996a). Recombination between sat-RNA D and TCV is therefore thought to be a very frequent event, resulting in populations of recombinants generated de novo in each infected cell that represent the original recombinant molecules rather than progeny of such molecules.

Most sat-RNA D/TCV recombinants recovered were the products of single recombination events; however, a few contained two or three discontinuous segments of TCV sequence, including one recombinant that contained sequence from both the plus and minus strands of TCV (CARPENTER and SIMON 1994). Crossover sites in TCV between positions 3800 and 3825 (3′ of the coat protein ORF and about 250 bases from the 3′ end) were found in most of the sat-RNA D/TCV recombinants. Two imperfect 24-base tandem repeats (motif 3A and motif 3B) that are similar in sequence to the 5′ ends of the subgenomic RNAs and can fold into a stable hairpin (motif3-hairpin) are contained in this hot-spot region. The major crossover site in this region is located just upstream of the motif3-hairpin, and the hairpin is required for recombination at the hot-spot position in vivo (CARPENTER et al. 1995). A role for the motif 3A/3B sequences in accumulation of TCV in plants and protoplasts was also found, as deletions that eliminated either motif and extended into the second either abolished or greatly decreased accumulation of TCV (CARPENTER et al. 1995). The motif3-hairpin was recently shown to be a transcription enhancer in vitro, similar to what was foundfor the motif1-hairpin (P.D. NAGY and A.E. SIMON, unpublished).

The locations of junction sites outside this TCV hot-spot region were dependent mainly on the age of the infection (CARPENTER and SIMON 1996a). This variation in populations of recombinants was seen in infected protoplasts and in whole plants. For example, early after inoculation of plants with TCV and sat-RNA D, recombinants with junctions outside the hot-spot region had crossover sites in TCV that were nearly all upstream of the motif3-hairpin, while several weeks later, nearly all such recombinants had junctions downstream of the motif3-hairpin. Since the TCV/sat-RNA D recombinants are "dead-end" products, i.e., cannot be amplified in cells, selection pressure to generate more fit recombinants apparently is not a factor in the accumulation of a particular population of recombinant molecules.

5 Repair of Sat-RNA 3′ End Truncations

RNA viruses with single-stranded 3′ ends that are not polyadenylated or amino-acylated may need to repair damage to single-stranded bases at their 3′ ends caused by cellular RNases. Therefore, the existence of a 3′-end repair mechanism (analogous to cellular telomerases) would be advantageous for RNA viruses. Plus-strand sat-RNAs and genomic RNA of TCV contain the motif 5′ CUGCCC-OH 3′ at their 3′ ends, which forms a single-stranded tail. Deletion of this sequence in sat-RNA C is repaired to the wild-type sequence in vivo (NAGY et al. 1997). The repair mechanism was shown to involve the 3′-end motif of the TCV genomic RNA, since alteration of the genomic RNA sequence from CCUGCCC to CCAGCCC resulted in repair of truncated sat-RNA C to the sequence CCAGCCC. The mechanism of repair was suggested to involve the TCV RdRp-mediated production of four- to eight-base oligoribonucleotides by abortive synthesis using the 3′ end of the TCV genomic RNA as template, a reaction that was shown to occur in vitro (NAGY et al. 1997). Such oligoribonucleotides were able to prime synthesis of wild-type minus-

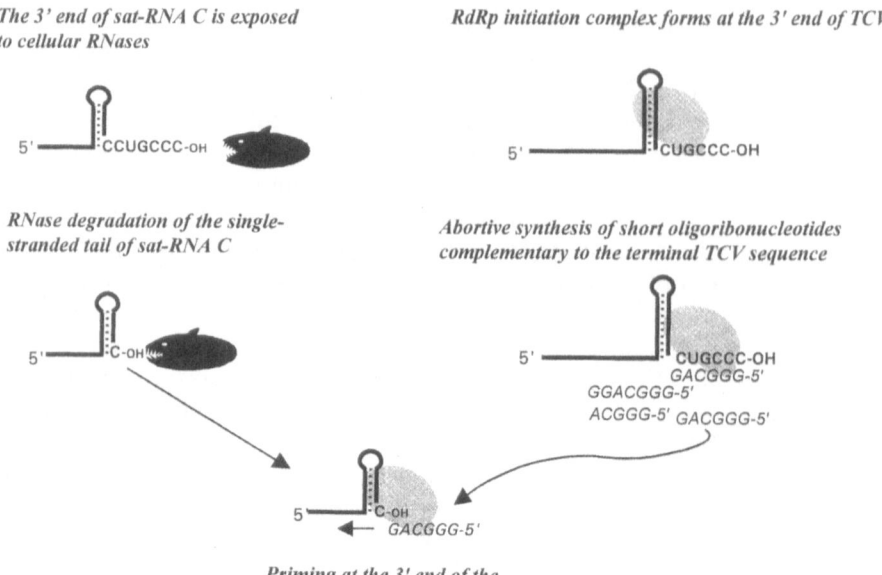

The 3' end of sat-RNA C is exposed to cellular RNases

RdRp initiation complex forms at the 3' end of TCV

RNase degradation of the single-stranded tail of sat-RNA C

Abortive synthesis of short oligoribonucleotides complementary to the terminal TCV sequence

Priming at the 3' end of the truncated sat-RNA C

Fig. 5. Model for repair of deletions at the 3' end of sat-RNA C plus strands. Priming at the 3' end of sat-RNA C using oligoribonucleotides generated from TCV requires a functional sat-RNA C hairpin sequence (NAGY et al. 1997). Synthesis of the oligoribonucleotide products does not occur using sat-RNA C templates that contain the 3'-terminal 100 nucleotides from TCV, indicating that inability of the RdRp to melt through the stable TCV 3'-end hairpin is not the major factor in generation of the abortive products

strand sat-RNA C in vitro without a requirement for base pairing of the oligoribonucleotides to the truncated, positive strand sat-RNA C template (Fig. 5).

Sat-RNA D transcripts with 3'-end deletions of five bases gave rise to wild-type sat-RNA in vivo, while deletions of 6–11 bases resulted in sat-RNA D with additional deletions to the –14 position joined to internal TCV genomic RNA (or other) sequence, followed by replacement of the terminal CCUGCCC motif (CARPENTER and SIMON 1996b,c). The selection of internal TCV genomic sequence used in the repair of sat-RNA D 3' ends was not random. Generation of these short RNA segments derived from internal TCV genomic RNA sequence likely involved primer-mediated synthesis of abortive products facilitated by base pairing between internal regions of the TCV genomic RNA and the oligoribonucleotides generated by abortive cycling from the 3' end of the TCV genome.

6 Origin of TCV Sat-RNAs

Since sat-RNA C is clearly derived from a rare double recombination event between sat-RNA D and the 3' end of TCV that led to the production of an infectious molecule, it is only the origin of sat-RNA D that is in question. Early work on the origin of TCV satellites suggested that the sat-RNAs originated from host DNA, since probes derived from sat-RNA C hybridized to several fragments in the turnip genome (ALTENBACH and HOWELL 1984). However, attempts to repeat such experiments were not successful, and no hybridization with full-length sat-RNA C probes was found using conditions where hybridization to single-copy genes should have been detectable (A.E. Simon, unpublished). A genomic origin for sat-RNA D was again postulated, based on the recovery of only wild-type sat-RNA D in all plants that were inoculated with mutant sat-RNA D transcripts (COLLMER et al. 1991). Interpreting the results of this study, however, is complicated by the unexplained appearance of sat-RNA C in plants that were not inoculated with this sat-RNA; the emergence of sat-RNA C was thought to have occurred by recombination between TCV and sat-RNA D that had reverted to wild type. Since the region shared by sat-RNA C and sat-RNA D differs in 21 positions, including seven insertions and deletions, and since over 1000 sequenced recombinants between sat-RNA D and TCV failed to reveal a molecule identical to sat-RNA C (CARPENTER et al. 1995), it is more likely that the appearance of sat-RNA C was due to contamination. This interpretation brings into question whether the reappearance of wild-type sat-RNA D in this study was also due to contamination. In the sat-RNA D 3'-end repair studies described above (CARPENTER and SIMON 1996b,c), numerous deletions in sat-RNA D were constructed and transcripts inoculated onto turnip plants; no wild-type sat-RNA D was recovered from these plants, with the exception of deletions of five bases or less at the 3' terminus.

If sat-RNA D did not arise from the genome of the host, what are other possible origins? As described in Sect. 5, repair of truncations at the 3' end of sat-RNA D probably involves two abortive cycling reactions using the TCV genomic RNA as template for one or both reactions (CARPENTER and SIMON 1996c). This results in the joining together of short segments derived from different regions of the TCV genome. One possibility is that sat-RNA D arose by a series of abortive cycling events and primer-extension reactions that joined discontiguous short segments of the TCV genome. If the resultant RNA was capable of being replicated and packaged, a new sat-RNA will have been "born". In support of this possibility was the finding that several 10- and 11-base segments of sat-RNA D are identical to sequence from the TCV genomic RNA (SIMON and NAGY 1996).

7 Prospects

Sat-RNAs provide an excellent model for studies on replication, recombination, and symptom production by viral RNAs. The small size, lack of open reading frames, and plasticity of many satellites have provided the means to make rapid advancements in our knowledge of many viral processes. TCV satellites have proven to be unusually amenable to such studies due to the availability of whole plant, protoplast, and in vitro systems to study replication and recombination. In addition, in vivo genetic selection can be used to analyze sequences that can function as RdRp promoters; such analyses have demonstrated that even the most simple RdRp promoter, such as the plus-strand promoter on sat-RNA C, can be very complex. Recent results have shown the interrelationship between RNA recombination and replication, with hairpins required for recombination in vivo serving as enhancers of transcription and requirements for primer extension in vitro. Such sat-RNA-based studies have proven for the first time that recombination is a process mediated by the viral RdRp.

The relationship among the host plant, virus genomic RNA, and sat-RNA in mediating symptoms remains an intriguing puzzle. Does the virus CP bind to the 3' end of sat-RNA C, and does this lead to symptom intensification? How does sat-RNA C affect the movement of the virus, enhancing movement into uninoculated emerging leaves in the presence of wild-type levels of CP and inhibiting movement when CP levels are reduced? What are the host factors involved in sat-RNA symptom mediation? The ability of TCV to infect *Arabidopsis* provides the opportunity to use genetic approaches to answer these questions. With such an abundance of questions remaining to be solved, the future of satellite research looks very bright indeed!

Acknowledgements. I am grateful to Drs. Peter Nagy and Clifford Carpenter for reviewing the manuscript. Work in my lab is supported by the National Science Foundation.

References

Altenbach SB, Howell SH (1981) Identification of a satellite RNA associated with turnip crinkle virus. Virology 112:25–33

Altenbach SB, Howell SH (1984) Nucleic acid species related to the satellite RNA of turnip crinkle virus in turnip plants and virus particles. Virology 134:72–77

Broadbent L, Heathcote GD (1958) Properties and host range of turnip crinkle rosette and yellow mosaic viruses. Ann Appl Biol 46:585–592

Carpenter CD, Simon AE (1994) Recombination between plus and minus strands of turnip crinkle virus. Virology 201:419–423

Carpenter CD, Simon AE (1996a) Changes in locations of crossover sites over time in de novo generated RNA recombinants. Virology 223:165–173

Carpenter CD, Simon AE (1996b) In vivo repair of deletions at the 3' end of a TCV satellite RNA in vivo may involve two abortive synthesis and priming events. Virology 226:153–160

Carpenter CD, Simon AE (1996c) In vivo restoration of biologically active 3' ends of virus-associated RNAs by non-homologous RNA recombination and replacement of a terminal motif. J Virol 70:478–486

Carpenter CD, Simon AE (1998) Analysis of sequences and putative structures required for viral satellite RNA accumulation by in vivo genetic selection. Nucleic Acids Res 26:2426–2432

Carpenter CD, Cascone PJ, Simon AE (1991) Formation of multimers of linear satellite RNAs. Virology 183:586–594

Carpenter CD, Oh J-W, Zhang C, Simon AE (1995) Involvement of a stem-loop structure in the location of junction sites in viral RNA recombination. J Mol Biol 245:608–622

Carrington JC, Morris TJ, Stockey PG, Harrison SC (1987) Structure and assembly of turnip crinkle virus. IV. Analysis of the coat protein gene and implications of the subunit primary structure. J Mol Biol 194:265–276

Carrington JC, Heaton LA, Zuidema D, Hillman BI, Morris TJ (1989) The genome structure of turnip crinkle virus. Virology 170:219–226

Cascone PJ, Carpenter CD, Li XH, Simon AE (1990) Recombination between satellite RNAs of turnip crinkle virus. EMBO J 9:1709–1715

Cascone PJ, Haydar T, Simon AE (1993) Sequences and structures required for RNA recombination between virus-associated RNAs. Science 260:801–805

Collmer CW, Stenzler L, Fay N, Howell SH (1991) Nonmutant forms of the avirulent satellite D of turnip crinkle virus are produced following inoculation of plants with mutant forms synthesized in vitro. Virology 183:251–259

Collmer CW, Stenzler L, Chen X, Fay N, Hacker D, Howell SH (1992) Single amino acid change in the helicase domain of the putative RNA replicase of turnip crinkle virus alters symptom intensification by virulent satellites. Proc Natl Acad Sci USA 89:309–313

Forster AC, Symons RH (1987) Self-cleavage of plus and minus RNAs of a virusoid and a structural model for the active sites. Cell 49:211–220

Gold L, Polisky B, Uhlenbeck O (1995) Diversity of oligonucleotide functions. Annu Rev Biochem 64:763–797

Guan H, Song C, Simon AE (1997) RNA promoters located on (–)-strands of a subviral RNA associated with turnip crinkle virus. RNA 1401–1412

Hacker DL, Petty ITD, Wei N, Morris TJ (1992) Turnip crinkle virus genes required for RNA replication and virus movement. Virology 186:1–8

Hogle JM, Maeda A, Harrison SC (1986) Structure and assembly of turnip crinkle virus. I. X-ray crystallographic structure analysis at 3.2A resolution. J Mol Biol 191:625–638

Kong Q, Oh J-W, Simon AE (1995) Symptom attenuation by a normally virulent satellite RNA of turnip crinkle virus is associated with the coat protein open reading frame. Plant Cell 7:1625–1634

Kong Q, Oh J-W, Carpenter CD, Simon AE (1997a) The coat protein of turnip crinkle virus is involved in subviral RNA-mediated symptom modulation and accumulation. Virology 238:478–485

Kong Q, Wang J, Simon AE (1997b) Satellite RNA-mediated resistance to turnip crinkle virus in Arabidopsis involves a reduction in virus movement. Plant Cell 9:2051–2063

Laakso MM, Heaton LA (1993) Asp → Asn substitutions in the putative calcium-binding site of the turnip crinkle virus coat protein affect virus movement in plants. Virology 197:774–777

Li XH, Simon AE (1990) Symptom intensification on cruciferous hosts by the virulent sat-RNA of turnip crinkle virus. Phytopathology 80:238–242

Li XH, Simon AE (1991) In vivo accumulation of a turnip crinkle virus DI RNA is affected by alterations in size and sequence. J Virol 65:4582–4590

Li XH, Heaton L, Morris TJ, Simon AE (1989) Defective interfering RNAs of turnip crinkle virus intensify viral symptoms and are generated de novo. Proc Natl Acad Sci U S A 86:9173–9177

Nagy PD, Simon AE (1997) New insights into the mechanisms of RNA recombination. Virology 235:1–9

Nagy PD, Simon AE (1998) In vitro characterization of late steps of RNA recombination in turnip crinkle virus I: role of the motif1-hairpin structure. Virology (in press)

Nagy PD, Simon AE (1998) In vitro characterization of late steps of RNA recombination in turnip crinkle virus II: role of the priming stem and flanking sequences. Virology (in press)

Nagy PD, Carpenter CD, Simon AE (1997) A novel 3′ end repair mechanism in an RNA virus. Proc Natl Acad Sci USA 94:1113–1118

Nagy PD, Zhang C, Simon AE (1998) Dissecting RNA recombination in vitro: role of RNA sequences and the viral replicase. EMBO J 17:2392–2403

Oh J-W, Kong Q, Song C, Carpenter CD, Simon AE (1995) Open reading frames of turnip crinkle virus involved in satellite symptom expression and incompatibility with Arabidopsis thaliana ecotype Dijon. Mol Plant Microbe Interact 8:979–987

Prody GA, Bakos JT, Buzayan JM, Schneider IR, Bruening G (1986) Autolytic processing of dimeric plant virus satellite RNA. Science 231:1577–1580

Qu F, Morris TJ (1997) Encapsidation of turnip crinkle virus is defined by a specific packaging signal and RNA size. J Virol 71:1428–1435

Simon AE, Bujarski JJ (1994) RNA recombination and evolution in infected plants. Annu Rev Phytopathol 32:337–362

Simon AE, Howell SH (1986) The virulent satellite RNA of turnip crinkle virus has a major domain homologous to the 3′-end of the helper virus genome. EMBO J 5:3423–3428

Simon AE, Nagy PD (1996) Recombination in the turnip crinkle virus system. Semin Virol 7:373–379

Simon AE, Engel H, Johnson R, Howell SH (1988) Identification of determinants affecting virulence, RNA processing and infectivity in the virulent satellite of turnip crinkle virus. EMBO J 7:2645–2651

Simon AE, Engel H, Howell SH (1989) Turnip crinkle virus satellite domains involved in virulence and processing. In: Staskowitz B, Ahlquist P, Yoder O (eds) Molecular biology of plant pathogen interactions. Liss, New York, pp 217–227 (UCLA symposia on molecular and cellular biology. New series, vol 101)

Simon AE, Li XH, Lew J, Stange R, Zhang C, Polacco M, Carpenter CD (1992) Susceptibility and resistance of Arabidopsis thaliana to turnip crinkle virus. Mol Plant Microbe Interact 5:496–503

Song C, Simon AE (1994) RNA-dependent RNA polymerase from plants infected with turnip crinkle virus can transcribe (+)- and (−)-strands of virus-associated RNAs. Proc Natl Acad Sci USA 91:8792–8796

Song C, Simon AE (1995a) Requirement of a 3′-terminal stem-loop in in vitro transcription by an RNA-dependent RNA polymerase. J Mol Biol 254:6–14

Song C, Simon AE (1995b) Synthesis of novel products in vitro by an RNA-dependent RNA polymerase. J Virol 69:4020–4028

Stupina V, Simon AE (1997) Analysis in vivo of turnip crinkle virus satellite RNA C variants with mutations in the 3′ terminal minus strand promoter. Virology 238:470–477

Wang J, Simon AE (1997) Analysis of the two subgenomic RNA promoters for turnip crinkle virus in vivo and in vitro. Virology 232:174–186

Zhang C, Cascone PJ Simon AE (1991) Recombination between satellite and genomic RNAs of turnip crinkle virus. Virology 184:791–794

Zhang C, Simon AE (1994) Effect of template size on replication of defective interfering RNAs. J Virol 68: 8466–8469

Structure and Functional Relationships of Satellite RNAs of Cucumber Mosaic Virus

F. García-Arenal[1] and P. Palukaitis[2]

1 Introduction

The first association of a satellite RNA with cucumber mosaic virus (CMV) was established by J.M. KAPER and colleagues (1976), who also showed that CMV containing this satellite RNA induced a lethal necrosis in the tomato (KAPER and WATERWORTH 1977). This initial discovery followed from earlier work in France concerning the search for the etiology of a lethal tomato necrosis that had appeared

[1]Departamento de Biotecnología, E.T.S.I. Agrónomos, Universidad Politécnica de Madrid, Ciudad Universitaria, 28040 Madrid, Spain
[2]Department of Virology, Scottish Crop Research Institute, Invergowrie, Dundee DD2 5AS, UK
FG-A received a grant from the *Fundación José Antonio de Castro*, Madrid, Spain. Unpublished work in the lab of PP was funded in part by grant no. 91-37303-6426 from the USDA NRICGP, grant no. 95-33120-1876 from the USDA BRARGP, grant no. DMB-9106293 from the US National Science Foundation, and a grant-in-aid from the Scottish Office Agriculture, Environment and Fisheries Department.

in the French Alsace in 1972 (MARROU et al. 1973). By 1974, it was clear that CMV was associated with the disease (MARROU and DUTEIL 1974). The epidemic did not re-occur for some time, although CMV strains outside of France apparently also contained necrogenic satellite RNAs (KAPER and TOUSIGNANT 1977; KAPER and WATERWORTH 1977). Nonetheless, the reappearance of the tomato-necrosis epidemic in Italy (GALLITELLI et al. 1988), as well as in Japan (KOSAKA et al. 1989) and Spain (JORDA et al. 1992), led to the unequivocal establishment of CMV strains containing necrogenic satellite RNAs as the causal agents of tomato necrosis.

Not all satellite RNAs of CMV are necrogenic for tomato (GOULD et al. 1978; KAPER et al. 1981; GONSALVES et al. 1982; PALUKAITIS 1988), and satellites that are necrogenic for tomato as well as those that are not attenuate CMV-induced symptoms on other hosts (WATERWORTH et al. 1979; MOSSOP and FRANCKI 1979a). Other satellite RNAs were discovered which induced necrosis on tomato and a yellow-leaf chlorosis on tobacco (TAKANAMI 1981), white-leaf chlorosis on tomato only (GONSALVES et al. 1982), yellow-leaf chlorosis on tomato only (PALUKAITIS 1998), or yellow-leaf chlorosis on tobacco only (PALUKAITIS 1988; SLEAT and PALUKAITIS 1990a). While KAPER and TOUSIGNANT (1977, 1978) initially found necrogenic satellite RNAs associated with all the CMV strains they examined, MOSSOP and FRANCKI (1978, 1979a; personal communication) found satellite RNAs associated only with CMV strains they were propagating, but not with some of the same strains kept in long-term storage or propagated elsewhere. This suggests that the satellite RNAs may have contaminated various CMV strains during greenhouse propagation, especially in tobacco (PIAZZOLLA et al. 1982a; GARCÍA-LUQUE et al. 1984). This was also indicated when it was discovered that the R- and S-strains of CMV maintained elsewhere contained non-necrogenic rather than necrogenic satellite RNAs (JACQUEMOND and LOT 1981; R.I.B. FRANCKI, personal communication), while after further propagation, such strains were found to contain either necrogenic satellite RNAs (KAPER and TOUSIGNANT 1977) or a mixture of necrogenic and non-necrogenic satellite RNAs (AVILA-RINCON et al. 1986a). In addition, it was established that a mixture of necrogenic and non-necrogenic satellite RNAs would induce tomato necrosis, even when only 0.5% of the satellite RNA inoculum was a necrogenic satellite RNA (JACQUEMOND and LOT 1981). Moreover, satellite RNAs of CMV were shown to be unusually resistant to ribonuclease, both in vitro and in vivo (MOSSOP and FRANCKI 1978, 1979b; JACQUEMOND and LOT 1982). Thus it became clear that satellite RNAs were highly infectious and persistent, and that strains of CMV could easily become contaminated with satellite RNAs. However, since it has been well documented that some CMV strains naturally contain mixtures of satellite RNAs, and specific variants are selected during passage in different hosts (GARCÍA-LUQUE et al. 1984; AVILA-RINCON et al. 1986a; KAPER et al. 1988; PALUKAITIS 1988; KURATH and PALUKAITIS 1989a; ARANDA et al. 1993; PALUKAITIS and ROOSSINCK 1996), in many cases the appearance of a pathogenic satellite may reflect such selection.

The problems associated with contamination, especially by necrogenic satellite RNAs, led KAPER et al. (1986, 1988) to re-examine the kinetics of satellite RNA-induced pathology, and they established a rigid bioassay protocol. Based on the dilution end point of the inoculum and a definition of necrogenicity as the ability to induce a lethal systemic necrosis in tomato by 11–20 days post inoculation, these workers were able to differentiate contamination from the effects due to the inoculum. Unfortunately, there were unintended consequences of this rigid definition of tomato necrosis. One consequence was the polemic about whether the Y-sat-RNA of CMV is truly necrogenic on tomato, or whether tomato necrosis is caused by either a contaminating necrogenic satellite RNA (KAPER et al. 1986) or environmental factors (WU et al. 1993a; KAPER et al. 1995). Another consequence was that a survey of *Lycopersicon* accessions indicated that in only a few accessions was tomato necrosis induced (WHITE and KAPER 1987). However, subsequent analysis of the accessions that did not exhibit a lethal systemic necrosis showed that they did exhibit necrosis to varying extents, making further genetic dissection of the necrotic response very complex (J. ABAD and F. GARCÍA-Arenal, unpublished data; D. BAULCOMBE, personal communication). In addition, Y-sat-RNA derived from a cDNA clone induced necrosis to different extents or not at all, depending on the cultivar of tomato and strain of helper virus (MASUTA et al. 1988a; MUSUTA and TAKANAMI 1989; WU et al. 1993a), while the Tfn-sat-RNA induced necrosis in tomato fruit but had no such effects on the foliage (CRESCENZI et al. 1993a). Thus, the rigid bioassay for necrogenicity could not be extended to other satellite RNAs, to many other tomato cultivars, or to in vitro generated mutants of satellite RNAs, where the mutagenesis might affect either the pathogenicity or the specific infectivity of that satellite RNA.

In spite of the technical difficulties indicated above, the large number of naturally occurring satellite RNAs of CMV have proven very useful for studies involving RNA sequence diversity, virus evolution, secondary structure determination, and RNA-mediated pathogenicity. In addition, the CMV satellite RNA system is a good model for virus replication, since it does not encode any factors required for its replication. Here, we will review the data and conclusions obtained under these various topics for the satellite RNAs of CMV.

2 Nucleotide Sequence and Structure

2.1 Nucleotide Sequence

Since RICHARDS et al. (1978) first reported the complete nucleotide sequence for a CMV satellite RNA 20 years ago, much work has been done on the molecular characterization of CMV satellite RNA variants, and today more than 100 sequences are reported in the literature or the data bases. They represent satellite RNAs originally found to be associated with about 65 CMV isolates or strains, coming from different geographical locations all over the world, and from different host plants. A large percentage of reported satellite RNA sequences correspond to different variants present in a single field isolate of CMV, or generated upon passage under controlled conditions (examples are listed in Fig. 1 and references therein).

Reported sequences for the linear, single-stranded, CMV satellite RNA molecules fall into two size classes: Most of them are of 332–342 nucleotide residues; satellite RNAs in this size class have been found in CMV isolates from all over the world and from many different host plant species and families. A limited number of reported variants are considerably larger, having 368–405 nucleotides. These variants have been found associated with only six CMV isolates from solanaceous host plants from Japan (four), China (one), or Italy (one) (HIDAKA et al. 1984, 1988; CRESCENZI et al. 1992; SAYAMA et al. 1993; HIDAKA and HANADA 1994). Interestingly, these larger satellite RNAs are very similar in sequence to those of the smaller size class, and sequence alignment shows that most of the sequences of the smaller satellite RNAs are also present in the larger ones. The difference in size can be attributed mostly to the insertion of new sequences relative to the smaller satellite RNA. These sequence insertions occur at specific sites in the smaller satellite RNAs. In four large satellite RNA variants [Y-, OY2-, and KN-sat-RNAs from Japan (HIDAKA et al. 1984, 1988; HIDAKA and HANADA 1994) and Tfn-sat-RNA from Italy (CRESCENZI et al. 1992)], the insertion substitutes for nucleotides 155–170 (compared with the satellite RNA 1-CARNA5; COLLMER et al. 1983). The inserted sequence in OY2- and Tfn-sat-RNA is virtually the same, and the inserted sequence in Y-sat-RNA and KN-sat-RNA is also the same, but different from that of OY2- and Tfn-sat-RNAs. No viral or host plant sequences have been identified similar to those inserted sequences, except for the insert in Y-sat-RNA, which has been reported to show significant complementarity to the sequence of a chloroplast tRNA gene (MASUTA et al. 1992). It should be stressed that the same insert has occurred in two satellite RNAs from very different regions of the world, and that these four satellite RNAs differ considerably in the rest of their sequence. KN-sat RNA has a second insertion that occurs between nucleotides 91 and 92 of the satellite RNA 1-CARNA5, and this insertion is also present in the satellite RNA 57-sat-RNA (SAYAMA et al. 1993). Again, the sequences of these insertions have not been identified as similar to other reported nucleotide sequences. Thus the acquisition of 'foreign' sequences is apparently limited to certain such sequences, occurs at determined sites in the satellite RNA molecule, and proceeds modularly.

All reported CMV satellite RNAs are linear, single-stranded RNA molecules. They all share a common decanucleotide sequence at the 5' end. For a number of satellite RNAs the 5' end has been reported to be capped, and this probably is the case for all of them (ROOSSINCK et al. 1992). The 3' end is also conserved between all reported satellite RNA variants in the last eight nucleotides. Comparison of 25 variants of CMV satellite RNA (FRAILE and GARCÍA-ARENAL 1991) showed polymorphic positions to be distributed uniformly over the molecule, with short conserved regions no longer than ten nucleotides. A comparison of the sequence of these 25 variants, excluding the inserted regions of the larger satellite RNAs, showed that sequence identity between any two variants ranged between 72.8 and 98.8%. Genetic distances, expressed as nucleotide substitutions per site and estimated according to the JUKES and CANTOR (1969) method, which takes into ac-

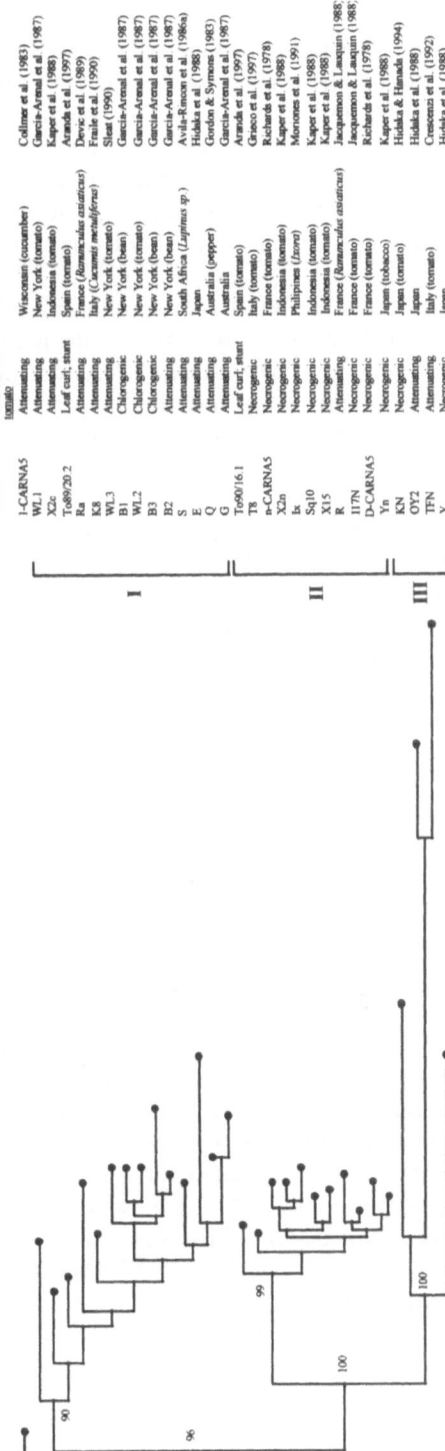

Set RNA	Symptoms on tomato	Origin (host)	Reference for sequence
I-CARNA5	Attenuating	Wisconsin (cucumber)	Collmer et al. (1983)
WL1	Attenuating	New York (tomato)	Garcia-Arenal et al. (1987)
X2c	Attenuating	Indonesia (tomato)	Kaper et al (1988)
To89/20 2	Leaf curl, stunt	Spain (tomato)	Aranda et al. (1997)
Ra	Attenuating	France (Ranunculus metalliferus)	Devic et al. (1989)
K8	Attenuating	Italy (Cucumis metalliferus)	Finale et al. (1990)
WL3	Attenuating	New York (tomato)	Sleat (1990)
B1	Chlorogenic	New York (tomato)	Garcia-Arenal et al (1987)
WL2	Chlorogenic	New York (tomato)	Garcia-Arenal et al (1987)
B3	Attenuating	New York (bean)	Garcia-Arenal et al (1987)
B2	Attenuating	New York (bean)	Garcia-Arenal et al (1987)
S	Attenuating	South Africa (Lupinus sp.)	Avila-Rincon et al (1986a)
E	Attenuating	Japan	Hidaka et al (1988)
Q	Attenuating	Australia (pepper)	Gordon & Symons (1983)
G	Attenuating	Australia	Garcia-Arenal et al (1987)
To60/16.1	Leaf curl, stunt	Spain (tomato)	Aranda et al (1997)
T8	Necrogenic	Italy (tomato)	Greaco et al (1997)
n-CARNA5	Necrogenic	France (tomato)	Richards et al (1978)
X2n	Necrogenic	Indonesia (tomato)	Kaper et al (1983)
Ix	Necrogenic	Philippines (ixora)	Moriones et al (1991)
Sq10	Necrogenic	Indonesia (tomato)	Kaper et al (1983)
X15	Necrogenic	Indonesia (tomato)	Kaper et al (1983)
R	Attenuating	France (Ranunculus asiaticus)	Jacquemond & Lauquin (1988;
117N	Necrogenic	France (tomato)	Jacquemond & Lauquin (1988;
D-CARNA5	Necrogenic	France (tomato)	Richards et al (1978)
Yn	Necrogenic	Japan (tobacco)	Kaper et al (1988)
KN	Attenuating	Japan (tomato)	Hidaka & Hanada (1994)
OY2	Attenuating	Japan	Hidaka et al (1988)
TFN	Attenuating	Italy (tomato)	Crescenzi et al (1992)
Y	Necrogenic	Japan	Hidaka et al (1988)

Fig. 1. Most parsimonious phylogenetic tree for 30 CMV satellite RNAs. *Numbers* at nodes indicate significance in a bootstrap analysis. Lengths of branches from the nodes are proportional to the number of steps in the parsimony analysis. Properties and origin of the satellite RNAs are also indicated. The following sets of satellite RNAs derive from the same field isolate of CMV:(R. Ra). (WL1. WL2. WL3). (B1. B2. B3). (X2c. X2n. X15. Sq10). (D-CARNA5. n-CARNA5). (Y. Yn)

count the possibility of multiple substitution and reversions per site, averaged 0.130 (range 0.338–0.012). These values were smaller when the larger satellite RNAs were excluded from the comparison (0.109 on the average, ranging from 0.192 to 0.012). This indicates that even in the regions conserved between the larger and smaller satellite RNAs, the larger satellite RNAs are more divergent than the smaller ones. This in turn suggests that the insertion of foreign sequences in these large satellite RNAs has an effect in promoting variation or in eliminating constraints to variation in the rest of the molecule.

Phylogenetic analyses of CMV satellite RNAs, both using parsimony or distance matrix methods, showed them to group into three main evolutionary lines (Fig. 1). Interestingly, the clustering of CMV satellite RNAs was unrelated to their geographical origin, no matter what scale was considered. Also, no correlation with host plant or with the subgroup of CMV strain (not shown) was found. There was also no strict correlation between the clustering of the satellite RNA variants and their phenotype, notably satellite RNAs attenuating symptoms on tomato appeared in all three clusters. Nevertheless, no satellite RNA variant necrogenic for tomato appeared in cluster I, and cluster II was formed mostly of variants necrogenic for this host plant. Genetic distances estimated separately for each of the three clusters formed by the 30 satellite RNA variants in Fig. 1 showed that the values are about three times larger for cluster III than for cluster I, and about 3.5 times larger for cluster I than for cluster II (values are 0.220, 0.075, and 0.022, respectively), indicating different degrees of constraint in the divergence of these three evolutionary lines.

Different open reading frames (ORFs) were found in the sequences of CMV satellite RNAs, and in vitro (but never in vivo) translation products have been reported for some of them (AVILA-RINCON et al. 1986b; HIDAKA et al. 1988, 1990). No ORF was universally conserved in all sequenced satellite RNAs, which suggests that they do not encode any product essential for their propagation and movement. Also, there was no correlation between the pathogenic phenotype of CMV satellite RNAs and the presence of any ORF (DEVIC et al. 1990; JAEGLE et al. 1990; MASUTA and TAKANAMI 1989). Sequence analysis has also provided evidence that these ORFs are not functional. An analysis of the mutations that occurred in 25 satellite RNA variants relative to a putative ancestral sequence showed that a large fraction (16.5%) of the total mutations were point insertions and deletions (FRAILE and GARCÍA-ARENAL 1991), a feature uncharacteristic of coding sequences (LI et al. 1981). Also, genetic distances per site calculated for the first, second, and third position of codons in the putative ORFs had similar values; this does not correspond to the pattern found in functional ORFs, where the value for the third position is higher than for the first or second (LI et al. 1981). These data indicate that CMV satellite RNA is not a coding RNA, and thus that the preservation of encoding sequences is not a constraint to its genetic divergence.

2.2 Structure

If CMV satellite RNAs are noncoding RNAs, their biological properties must depend on the direct interaction between the satellite RNA and components of the helper virus and/or host plant (GARCÍA-ARENAL et al. 1987; WU and KAPER 1992; BERNAL and GARCÍA-ARENAL 1994a). Thus, the secondary and higher-order structure of CMV satellite RNAs must be important for their biological properties, as has been shown for several noncoding, functional RNAs, including other viral and subviral plant pathogens (BRUENING 1990; FELDEN et al. 1994; OWENS et al. 1996).

Several analyses have been made of the structure of CMV satellite RNA variants, and secondary structure models have been proposed. The first such model was proposed by GORDON and SYMONS (1983) for Q-sat-RNA, a variant belonging to cluster I in Fig. 1. The proposed model was based on the identification of unpaired nucleotide residues by their susceptibility to nuclease S1 and to ribonuclease T1 (RNase T1). Also, direct information on base-paired regions was obtained by the analysis of paired fragments after partial enzymatic digestion and a two-step gel fractionation method. In this model 52% of the 336 nucleotides of Q-sat-RNA are involved in 87 base pairs (47 G:C, 24 A:U, 16 G:U pairs). The model (Fig. 2A) showed several structural elements involving the 5′ third and 3′ third of the molecule, with little structure in the central third, where alternative configurations are possible. A configuration reminiscent of a truncated tRNA-like structure was proposed for the very 3′ end (GORDON and SYMONS 1983). The fit of the experimental data to the model is generally good, particularly in the elements involving the 5′ third of the satellite RNA. This model was later extended to four other satellite RNAs in cluster I (B2- B3-, G- and WL1-sat-RNAs) , based on the mapping of unpaired residues using nuclease S1 and RNase T1, and in one case using adenine modification by diethyl pyrocarbonate (DEP) (GARCÍA-ARENAL et al. 1987). Although these four satellite RNAs could be folded into the same structure as Q-sat-RNA, the fit of the experimental data was not as good, and varied considerably for the different variants. Again, the fit was best in the 5′ part of the molecule. Interestingly, variations in the pattern of enzymatic cleavage was found for B2- and B3-sat-RNAs in regions of identical sequence. HIDAKA et al. (1988) have presented data for accessibility of phosphodiester bonds to cleavage by nuclease S1 and RNase T1 for E-sat-RNA, in cluster I and for Y- and OY2-sat-RNA, in cluster III. The data also fit the model of GORDON and SYMONS (1983) in the 5′ part, and include a less structured central part. Thus, several main features of the model proposed by GORDON and SYMONS (1983) appear to be of general application, including some structural elements, notably stem-loop I (Fig. 2).

The secondary structure of Ix- and D4-sat-RNAs, in cluster II of Fig. 1, was analyzed recently (BERNAL and GARCÍA-ARENAL 1997). Unpaired nucleotides were mapped by their susceptibility to cleavage by nuclease S1 and RNase T1, and to modification by dimethylsulfate (DMS), and unstacked adenines were mapped by their susceptibility to DEP. Information on base-paired positions was obtained by cleavage with nuclease VI and by the analysis of base-paired fragments as in GORDON and SYMONS (1983). Direct information was thus obtained for 90% of the

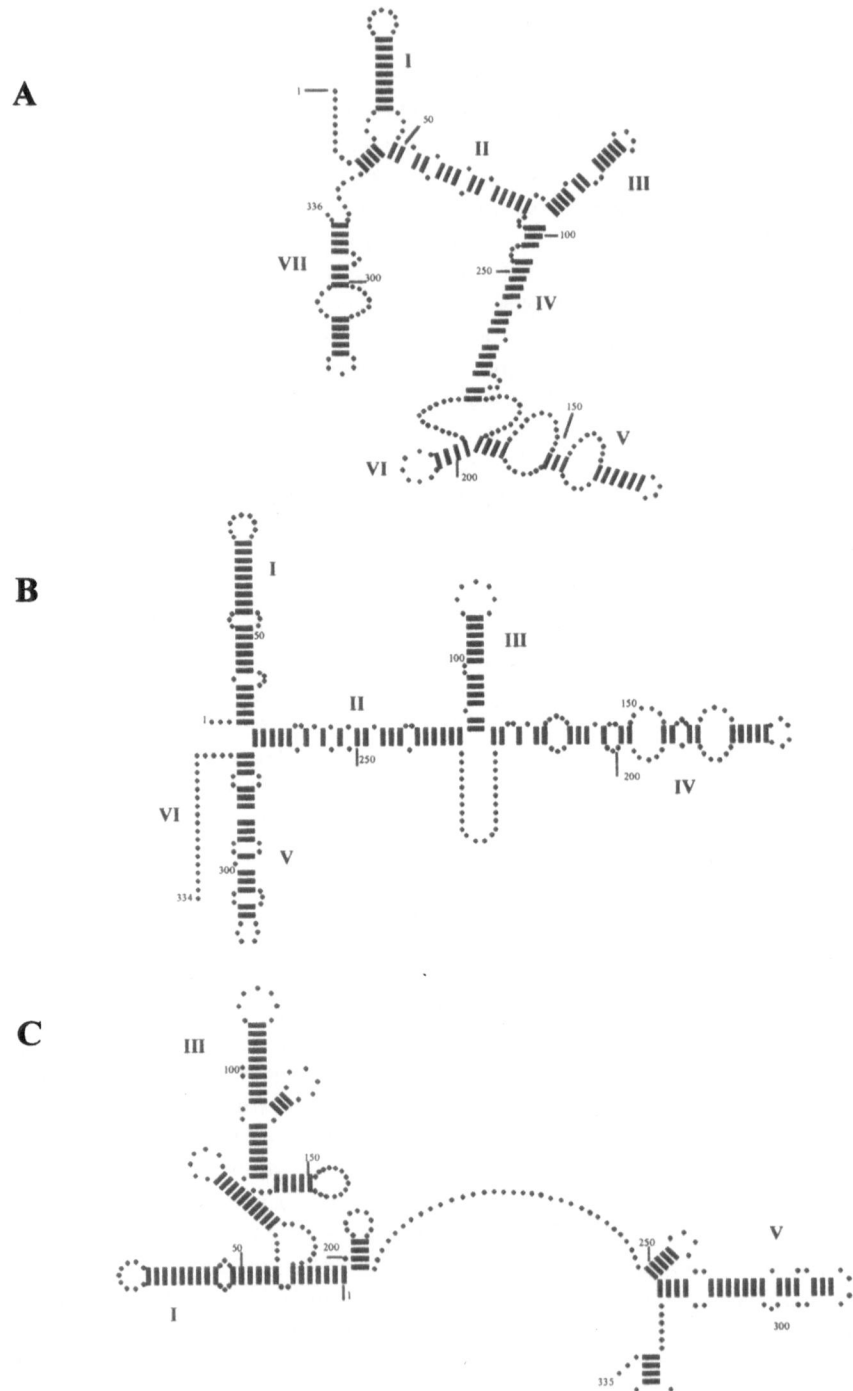

Fig. 2A–C. Models proposed for the in vitro secondary structure of Q-sat-RNA (**A**, GORDON and SYMONS 1983), Ix5-sat-RNA (**B**, BERNAL and GARCÍA-ARENAL 1997), and D4-sat-RNA (**C**, RODRÍGUEZ-ALVARADO and ROOSSINCK 1997)

334–335 positions of these satellite RNAs, the most detailed structural mapping reported for a CMV satellite RNA. In the proposed secondary structure model (BERNAL and GARCÍA-ARENAL 1997), 51% of the nucleotides were arranged in 86 pairs (40 G:C, 18 A:U, 28 G:U). The fit of the experimental data to the model is generally good, particularly so in the more structured elements in the 5′ and 3′ regions (elements I, II, III and V, Fig. 2B). As in the model of GORDON and SYMONS (1983), secondary structure elements were concentrated in the 5′ third and 3′ third of the molecule, while little structure was found in the central region (IV in Fig. 2B). On the other hand, only stem loops I and V of the model of GORDON and SYMONS (1983) (although with a longer stem in I) occur in the configuration of BERNAL and GARCÍA-ARENAL (1997) (as stem-loops I and IV, respectively), with the well-documented stem loops II and V having no counterpart in the model of GORDON and SYMONS (1983). No satisfactory structure was found for the 3′ terminus, but a truncated tRNA-like structure was incompatible with the experimental data. In addition, DMS analyses of six satellite RNA variants that are recombinants between Ix- and D4-sat-RNAs showed that sequence variation may result in different patterns of DMS modification in regions were the sequences are identical. These data suggest that tertiary interactions could occur between some of the structural elements in Fig. 2B, but no direct analysis of tertiary structure was done. In any case, no evidence for pseudoknots, a common motif of tertiary structure in RNAs (PLEIJ et al. 1987) was found.

All the previous reports addressed only the analysis of the in vitro secondary structure of CMV satellite RNA. For D4-sat-RNA, RODRIGUEZ-ALVARADO and ROOSSINCK (1997) analyzed the susceptibility of adenines and cytosines to DMS modification in vitro, in vivo, and in virions, and based on these data plus computer analyses, three secondary structure models were proposed. The model proposed for the in vitro structure (Fig. 2C) shares with the one proposed by BERNAL and GARCÍA-ARENAL (1997) the presence of stem-loops I, III, and V and a poorly structured central part. Interestingly, the accessibility of adenines and cytosines to DMS varied for this central part of the molecule, when the modification was done in vivo vs. in vitro vs. in virions, indicating alternative structures for this molecular region in the three different situations analyzed. The terminal part of stem-loop V is a conserved motif in satellite RNA variants necrogenic for tomato, which cannot be formed in non-necrogenic satellite RNAs. Thus, it was speculated that this structural element might be related to pathogenesis (RODRIGUEZ-ALVARADO and ROOSSINCK 1997).

The data reported in the studies discussed above show that some features are common to all proposed models. One main common feature is that the positions that would be based paired are mostly the same in all three models of Fig. 2, as already discussed for models *A* and *C*. Also, satellite RNA variants in clusters I and II of Fig. 1 can be folded according to the structures shown in Fig. 2A and B (the analysis has not been done with the model in Fig. 2C). Thus, it is possible that a general structure common to all satellite RNAs in the 330–340 size class could exist. Also, it is possible that the structure of satellite RNAs in clusters I and II could differ. Since the proposed models are based on experimental data of very different degrees of precision, these alternative possibilities cannot be resolved at present.

The close similarity between the sequences of CMV satellite RNAs does not allow the use of the phylogenetic comparative method to infer a secondary structure and show its conservation during evolution (FRAILE and GARCÍA-ARENAL 1991). However, comparative phylogenetic analyses revealed covariation for some positions: two of them occurred in the distal part of stem-loop I and supported the presence of this structural element, as appears in all three models in Fig. 2; the third one was common to all satellite RNAs necrogenic for tomato, in stem loop V of Fig. 2B and C. The pattern of mutations present among 25 satellite RNA variants was shown to be different for positions that would be base paired, or unpaired, according to the model of GORDON and SYMONS (1983; FRAILE and GARCÍA-ARENAL 1991). The total number of point mutations, the number of point insertions and deletions, and the relative number of transversions versus transitions were shown to be significantly higher for unpaired than for paired positions. The same trend was found on the analysis of 23 satellite RNA variants from field epidemics (ARANDA et al. 1997). These data support the general validity of the secondary structure model by GORDON and SYMONS (1983). Nevertheless, since the base-paired positions in the models of GORDON and SYMONS (1983) and BERNAL and GARCÍA-ARENAL (1997) were mostly the same (i.e., 80% of the residues that are base paired in one model are base paired in the other; BERNAL and GARCÍA-ARENAL 1997), this analysis supports models *A* and *B* in Fig. 2 equally. In any case, what this analysis of the pattern of mutations showed clearly was that the maintenance of a functional structure is a constraint to the genetic divergence of CMV satellite RNAs.

3 Genetic Variability Under Experimental and Natural Field Conditions

The genetic heterogeneity of CMV satellite RNA variants associated with CMV isolates was already shown by RICHARDS et al. (1978). When those authors first determined the sequence of the satellite RNA associated with strain D of CMV, they identified a number of closely related sequence variants. The genetic variability of the satellite RNA sequence population associated with a CMV isolate was further shown by the separation of different sequence variants upon passaging in different host plant species. GARCÍA-LUQUE et al. (1984) and KAPER et al. (1988) showed that satellite RNA variants necrogenic as well as non-necrogenic for tomato were present in a CMV isolate, and that different variants were selected for upon passaging in different host plant species. A similar observation was made with chlorosis-inducing satellite RNAs (PALUKAITIS 1988). At that time, it had already been shown that RNA populations, including those of plant RNA viruses, were built of different sequences which varied around an average, or master sequence, a population structure that has been called a quasispecies (DOMINGO et al. 1978; RODRIGUEZ-CEREZO and GARCÍA-ARENAL 1989; see DOMINGO and HOLLAND 1997 for a recent review). A detailed analysis of satellite RNA population structure done by KURATH and PALUKAITIS (1989a), showed that different cDNA clones from each of three different satellite RNA isolates differed in sequence according to a quasi-species distribution, and that the master sequence of the quasispecies could shift upon passage in different host plants. Those authors further showed that CMV satellite RNA populations starting from RNA transcripts of full-length cDNA clones, i.e., from a homogeneous RNA population, acquired upon multiplication *in planta* a heterogeneous, quasispecies structure, and that the master sequence of this population could change upon passage (KURATH and PALUKAITIS 1990).

It is important to understand which factors determine the changes in the prevalent sequence of a satellite RNA population, as this will be relevant to understanding the evolution of CMV satellite RNA populations. Evolution of satellite RNA populations is obviously of basic interest for the understanding of RNA virus evolution, but is also very important with regard to the possible use of CMV satellite RNA as a biocontrol agent, in transgenic or in cross-protection programs, of CMV-induced diseases (TIEN and WU 1991). Much work has been done in this direction in different laboratories. A main factor evident in many reports is the random effect of genetic drift, associated with population bottle necks during transmission (GARCíA-LUQUE et al. 1984; KAPER et al. 1988; KURATH and PALUKAITIS 1989a, 1990). Although the size of these bottlenecks has not been estimated, the effect of random genetic drift will possibly be much more important in natural aphid transmissions than in experimental transmissions with purified virus or with sap from infected plants. The species of host plant may also play a major role in determining the outcome of satellite RNA evolution, and the consistent selection of different sequence variants in different host species has been shown repeatedly (KAPER et al. 1988; KURATH and PALUKAITIS 1990; MORIONES et al. 1991; SMITH et al. 1992). A second major factor is the nature of the helper virus, which has also been shown to consistently select certain sequence variants from heterogeneous mixtures (PALUKAITIS and ROOSSINCK 1995; ROOSSINCK and PALUKAITIS 1995). In some cases, the regions of the satellite RNA molecule that determine an increased fitness in the presence of a certain helper virus have been mapped; these regions are not the same for different CMV satellite RNAs (JAEGLE et al. 1990; ROOSSINCK and PALUKAITIS 1995). Finally, the nature of the sequence variant itself may determine the outcome of satellite RNA evolution, since it has been shown that this is dependent on the initial sequence context (PALUKAITIS and ROOSSINCK 1995). Less information is available on the influence of other environmental factors, although temperature has also been shown to play a role in the outcome of satellite RNA population evolution (KAPER et al. 1995; WHITE et al. 1995). It is important to stress that under the same circumstances the outcome of evolution may be the same (i.e., the same sequence variant is selected for), and that this may lead to important phenotypic changes on the satellite RNA, such as the change from a benign to a pathogenic form (PALUKAITIS and ROOSSINCK 1996).

Although CMV satellite RNAs have been reported in many different CMV isolates from all over the world, the frequency of satellite RNAs in field populations of CMV seems to be usually low (KEARNEY et al. 1990). Work on the evolution of satellite RNA populations under natural, field conditions derives mainly from the analysis of populations representing epidemics of tomato necrosis caused by

CMV+ satellite RNA that occurred in southeastern Spain and southern Italy in the late 1980s. In the field, CMV satellite RNA populations are highly heterogeneous, much more so than the corresponding populations of the helper virus (ARANDA et al. 1993; FRAILE et al. 1997). Genetic heterogeneity seems to develop quickly by the sequential accumulation of point mutations (ARANDA et al. 1993; GRIECO et al. 1997). There is evidence of an upper threshold for genetic divergence, indicating that evolutionary constraints operate (ARANDA et al. 1993; GRIECO et al. 1997), at least in part, due to the maintenance of a functional structure (ARANDA et al. 1997). The insertion of foreign sequences to generate larger satellite RNAs, which has been shown to occur in the Italian population (CRESCENZI et al. 1992), may be a mechanism of escape from these constraints. Analysis of the evolutionary dynamics of field populations showed that satellite RNAs necrogenic or non-necrogenic on tomato evolved separately, clustering into two main evolutionary lines (ARANDA et al. 1997; GRIECO et al. 1997), but recombination between types from both lines may be frequent, disrupting this main evolutionary pattern (ARANDA et al. 1997). In contrast to what is apparent in experiments under controlled conditions, neither the host plant nor the strain of the helper virus had a major influence in the evolution of CMV satellite RNA in the field: variants from the analyzed epidemics did not cluster according to the host plant or according to the strain of helper virus (GRIECO et al. 1997; ALONSO-PRADOS et al. 1998). It is possible that other, unidentified factors are more important in field conditions as determinants of CMV satellite RNA evolution. It is also possible that the genetic bottlenecks associated with aphid transmission from plant to plant introduce a random element that overshadows any effect of host- or helper-associated selection. Data from Spain also showed that satellite RNA sampled from six different areas over 8 years belonged to a single population that was not differentiated according to place or year (ALONSO-PRADOS et al. 1998), which was in sharp contrast to the metapopulation structure of the helper virus, CMV (FRAILE et al. 1997). This showed that the population dynamics and genetics of CMV and its satellite RNA are uncoupled, and that CMV satellite RNA spreads epidemically, as a parasite, on the population of its helper virus.

Thus, data from the analysis of natural populations showed that CMV satellite RNA behaves as a molecular parasite, highly variable because of mutation accumulation, recombination, and acquisition of foreign sequences, and that the evolution of these populations follows dynamics difficult to predict. These results raise concern about the risks associated with the widespread use of satellite RNA for the biocontrol of CMV (see Sect. 6).

4 Pathogenicity

Depending on the propagation host, the helper virus, and the particular satellite RNA, the replication of satellite RNA could have three types of effects on the pathogenicity of the helper virus: (a) the pathogenicity could be unaffected; (b) the pathogenicity could be enhanced, by the induction of either necrosis or chlorosis; or (c) the pathogenicity could be attenuated. In most situations (i.e., most satellite RNAs tested with a range of helper viruses on a number of host species), satellite RNAs attenuated the pathogenicity induced by the helper virus (Mossop and Francki 1979a; Waterworth et al. 1979; Jacquemond and Leroux 1982; Palukaitis 1988). No satellite RNA has yet been described that was incapable of attenuating the symptoms induced by some tested strains of helper virus on a range of host species. Thus, even if particular satellite RNA sequences are involved in attenuation of helper virus-induced pathogenicity, the nature of those particular sequences remains, as yet, unknown.

It has been suggested that the attenuation associated with satellite RNAs is due to their competition with CMV for replication (Kaper 1982), since the replication of satellite RNA correlated with the accumulation of high levels of double-stranded (ds) satellite RNA and a reduction in CMV RNA accumulation (Habili and Kaper 1981; Piazzolla et al. 1982b). However, when tomato aspermy cucumovirus (TAV) was the helper virus, some satellite RNAs were able to attenuate the symptoms induced by TAV but did not reduce the levels of accumulation of the TAV RNAs (Harrison et al. 1987; Moriones et al. 1992). In the latter instance, the ds satellite RNA did not accumulate to high levels (Moriones et al. 1992); this would still leave a role for the accumulated ds satellite RNA itself in attenuation of viral symptoms (see Sect. 5.2).

The localization of sequences specifying the phenotypes associated with particular satellite RNAs of CMV, i.e., either necrosis in tomato or chlorosis in tomato or tobacco, was possible once infectious RNA transcripts were generated from cDNA clones of individual satellite RNAs. By generating chimeric satellite RNAs between pathogenic and non-pathogenic CMV satellite cDNA clones, sequences specifying chlorosis vs. necrosis were delimited to the 5′ half and 3′ half of the satellite RNA molecules, respectively (Devic et al. 1989; Kurath and Palukaitis 1989b; Masuta and Takanami 1989). Further delimitation to specific sequence domains was accomplished by mutagenesis of infectious cDNA clones. First, sequence comparisons among pathogenic and non-pathogenic RNAs were made to establish correlations between the presence of particular sequences and specific pathogenic responses; then site-directed mutagenesis was done to determine which of those alterations influenced pathogenicity (Masuta and Takanami 1989; Sleat and Palukaitis 1990b, 1992; Devic et al. 1990; Jaegle et al. 1990). These data established two pathogenicity domains (Fig. 3), the exact borders of which are not known but are inferred from sequence comparison data.

4.1 Necrosis Induction

In the case of necrogenic satellite RNAs, sequence changes outside the necrosis domain were found to influence the extent of necrosis, as well as the influence of helper virus on the induction of necrosis (Sleat et al. 1994; Wu and Kaper 1992). It was suggested that perhaps alteration in the secondary structure of the satellite

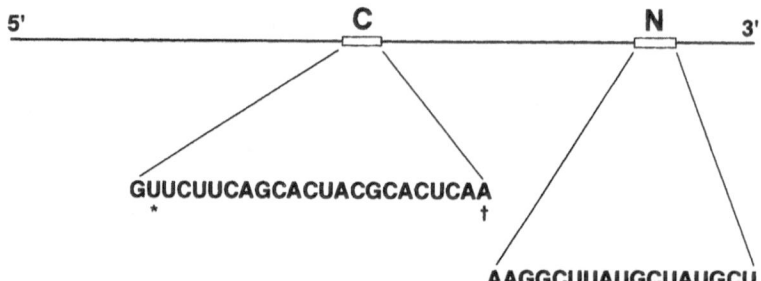

Fig. 3. Pathogenicity domains of CMV satellite RNAs. The chlorosis-inducing (*C*) and necrogenic (*N*) domains are shown in opposite halves of the satellite RNA molecule (of a standard 335-nt satellite RNA). The sequences of the *C* and *N* domains are shown, as well as sequences in the *C* domain that affect the host specificity for chlorosis induction (*, tobacco vs. tomato) or the extent of chlorosis in tomato (†, white vs. yellow leaves)

RNA might influence pathogenicity (SLEAT et al. 1994), and computer-generated secondary structures indicated that the necrogenicity domain might be located in the stem of a hairpin structure (TOUSIGNANT and KAPER 1993). However, such a stem-loop structure was not observed in experimentally determined secondary structures of necrogenic satellite RNAs (BERNAL and GARCíA-ARENAL 1997; RODRIGUEZ-ALVARADO and ROOSSINCK 1997), although another structure was found conserved in necrogenic satellite RNAs (RODRIGUEZ-ALVARADO and ROOSSINCK 1997).

Recently, a necrogenic satellite RNA of CMV was expressed in the potato virus X (PVX) vector and induced necrosis on the tomato, even when only the 3′ half of the necrogenic satellite RNA was expressed from the PVX vector (TALIANSKY et al. 1998). Curiously, necrosis was not observed when the (+) satellite RNA sequence (the polarity found encapsidated) was expressed in the same orientation as the (+) PVX RNA. Rather, necrosis was induced only when (−) satellite RNA sequences were expressed in the same orientation as (+) PVX RNA. These observations indicate that the (−) satellite RNA, rather than the (+) satellite RNA, contains the necrogenicity domain. Computer-generated secondary structures of the 5′ half of (−) satellite RNA (complementary to the 3′ half of the (+) satellite RNA, to which the necrogenicity domain had been mapped previously) were similar to those generated previously by TOUSIGNANT and KAPER (1993) for the complementary (+) satellite RNA sequences, with the necrogenicity domain located in the loop and adjacent stem of a hairpin structure. The corresponding sequence of a non-necrogenic satellite RNA contained a different sequence in the loop of a similar hairpin. The nature of the host factors involved in necrosis induction remains unknown.

4.2 Chlorosis Induction

In the case of chlorosis-inducing satellite RNAs, both the pathogenicity phenotype itself and the pathogenicity domain are more complex. First, for a given satellite

RNA chlorosis induction can occur in either tobacco or tomato, but not both (GONSALVES et al. 1982; PALUKAITIS 1998; DEVIC et al. 1989; KURATH and PALUKAITIS 1989b; MASUTA and TAKANAMI 1989; JAEGLE et al. 1990). Second, for some satellite RNAs the strain of helper virus can influence chlorosis induction; in one case, chlorosis induction in tobacco was induced in the presence of CMV strains in subgroup II, but not in subgroup I, and chlorosis induction mapped to RNA 2 of the subgroup II strain (SLEAT and PALUKAITIS 1990a). Third, chlorosis in tomato can be yellow (loss of chlorophyll) or white (loss of both chlorophyll and xanthophyll), depending on the satellite RNA and the particular helper virus strain (PALUKAITIS 1988; ZHANG et al. 1994). And fourth, chlorosis in tomato is a psychrogenic response, occurring at temperatures below 25°C and only in newly developing tissues (P. PALUKAITIS, unpublished observation). Mapping of the chlorosis-induction domain for several satellite RNAs delimited the same sequences and also identified a number of sequence residues within this domain that did not affect the phenotype (JAEGLE et al. 1990; KUWATA et al. 1991; SLEAT and PALUKAITIS 1992). In addition, specific sequence changes which affected the host specificity of chlorosis (tobacco vs. tomato; SLEAT and PALUKAITIS 1992) as well as the extent of chlorosis (yellow chlorosis vs. white chlorosis; SLEAT et al. 1994) were identified (Fig. 3). Since chlorosis induction was not due to expression of any open reading frames (MASUTA and TAKANAMI 1989), the data suggested that the RNA sequence itself was important in the induction of chlorosis.

4.3 Helper Virus and Host Genotype Effects

Expression of either the chlorosis-inducing Y-sat-RNA in transgenic tobacco (MASUTA et al. 1989) or the necrogenic D-sat-RNA in transgenic tomato (McGARVEY et al. 1990) did not induce pathogenicity. However, when those plants were challenged with CMV, the transgenically expressed satellite RNA was amplified and the pathogenic response associated with the particular satellite RNA ensued. Therefore, expression of (+) satellite RNA sequences in transgenic plants was not sufficient to induce pathogenicity.

The nature of any association between the chlorosis-inducing satellite RNAs, the helper virus, and specific host components resulting in induction of the various chlorosis responses remains unknown. However, chlorosis in tobacco apparently is not due to alterations in the lipid composition of cytoplasmic or chloroplast membranes (MASUTA et al. 1993a), even though significant nucleotide sequence complementarity was detected between a region of Y-sat-RNA containing the chlorosis domain and a chloroplast tRNA (MASUTA et al. 1992).

The chlorosis response induced by satellite RNA in tobacco and several other *Nicotiana* species was examined and was shown to be controlled by a single, incompletely dominant gene (MASUTA et al. 1993b). By contrast, chlorosis in tobacco induced by the Y-strain of CMV itself is elicited by the virus coat protein and is controlled through two recessive host genes (TAKAHASHI and EHARA 1993). No work on the genetics of the chlorosis response in tomato has been reported.

However, since PVX expressing a chlorosis-inducing satellite RNA (in either orientation) did not induce chlorosis in tomato (M. TALIANSKY, E.V. RYABOV, D.J. ROBINSON and P. PALUKAITIS, unpublished data), and the nature of chlorosis in tomato varied with the strain of helper virus (PALUKAITIS 1988; SLEAT and PALUKAITIS 1990a) as well as specific satellite RNA sequences (SLEAT et al. 1994), it is likely that several genes in tomato will influence the interactions leading to white-leaf or yellow-leaf chlorosis.

5 Replication

5.1 Helper Virus Effects on Replication

The satellites of CMV are dependent upon CMV for replication and encapsidation (GOULD et al. 1978; LINTHORST and KAPER 1985). Most strains of CMV examined can replicate and package various satellites of CMV, although there is one exception (KAPER et al. 1990). Most, but not all satellites of CMV are poorly supported in cucurbit hosts (KAPER and TOUSIGNANT 1977; JACQUEMOND and LEROUX 1982; PALUKAITIS 1988; MORIONES et al. 1991; ROOSSINCK and PALUKAITIS 1995). The inability of one strain of CMV to support satellite RNA in squash was at the level of satellite RNA replication, and mapped to CMV RNA 1 sequences adjacent to the putative helicase domain VI (ROOSSINCK et al. 1997). The high accumulation of the Ix-sat RNA did not map to one specific sequence, but rather appeared to be due to the conformation of the satellite RNA (BERNAL and GARCÍA-ARENAL 1994a).

Some strains of TAV were able to support a variety of satellite RNAs (GOULD et al. 1978; LEE and KUMMERT 1985; HARRISON et al. 1987), but other TAV strains did not support particular satellite RNAs, either for replication (JAEGLE et al. 1990) or for encapsidation and systemic movement (MORIONES et al. 1992). In the former case, it was found that the Y-sat-RNA was not supported by TAV unless sequences in both termini were exchanged with those of the R-sat-RNA (JAEGLE et al. 1990). In the latter case, TAV was able to replicate Ix-sat-RNA but was unable to either encapsidate or promote the long-distance movement of Ix-sat-RNA (MORIONES et al. 1992). The specific satellite sequences involved in the latter interactions could not be delimited; the data once again suggested that the interactions were affected by conformational changes in the satellite RNAs (BERNAL and GARCÍA-ARENAL 1994b). In the case of satellite RNA mixtures, TAV showed preferential support for D4-sat-RNA over WL47-sat-RNA, while the LS-strain of CMV supported WL47-sat-RNA over D4-sat-RNA (ROOSSINCK and PALUKAITIS 1995). From mixtures of reciprocal chimeras formed between cDNA clones of these two satellite RNAs, TAV preferentially accumulated satellite RNAs with satellite RNA sequences derived from the third quarter of the D4-sat-RNA molecule, while LS-CMV preferentially accumulated those satellite RNAs that contained the same domain from the WL47-sat-RNA (ROOSSINCK and PALUKAITIS 1995). Although in this situation it was not ascertained whether these differences in support were at the level of replication, encapsidation, or movement, it is clear from the three examples above that different sequences or conformational structures can affect the support of different satellite RNAs by TAV as well as by particular strains of CMV.

Pseudorecombinant viruses made between CMV and TAV, by exchange of RNA 3, were used to map the differential satellite support functions to RNAs 1 and/or 2, and not to the coat protein gene on RNA 3 (MORIONES et al. 1994). Thus, specific interactions with either RNAs 1 and 2 or their encoded genes (1a, 2a, 2b) are responsible for systemic movement of satellite RNAs.

5.2 Replication In Vivo

An early observation in plants infected by CMV containing satellite RNA was the high level of ds RNA that accumulated, over 90% of which was ds satellite RNA (DIAZ-RUIZ and KAPER 1977). In contrast to the situation observed with CMV alone (or other plant, positive-sense, RNA viruses), where the ratio of (+):(−) viral RNA was 20–100:1, the ratio of (+):(−) satellite RNA was 2–3:1 (PIAZZOLLA et al. 1982b), with most but not all of the (−) satellite RNA in a ds RNA form after extraction (HABILI and KAPER 1981; KURODA et al. 1997). A model for satellite RNA-mediated attenuation was proposed (KAPER 1982), based on the observation that symptomless leaves of infected tobacco plants contained high levels of ds satellite RNA and low levels of CMV, while the symptom-bearing leaves of the same infected tobacco plants contained high levels of CMV and single-stranded (ss) satellite RNA, but low levels of ds satellite RNA (HABILI and KAPER 1981). However, in the absence of satellite RNA, the same strain of CMV (P. Palukaitis, unpublished data), as well as other strains (GAL-ON et al. 1995), showed the same cycling of pathology and changes in the level of CMV RNA accumulation. Thus, the CMV cycling phenotype in tobacco also correlated with both an alteration in the ratio of ss to ds satellite RNA and synthesis of ds satellite RNA to high levels in leaves supporting low levels of CMV accumulation. Whether these two phenomena are caused by the same response is unknown.

Besides high levels of unit-length ds satellite RNAs, plants infected by CMV containing satellite RNA also produced a series of multimers (dimers, trimers, etc.) (LINTHORST and KAPER 1985; YOUNG et al. 1987; KURODA et al. 1997). Both (+) and (−) polarity multimers have been detected (KURODA et al. 1997; YOUNG et al. 1987), although no free (−) satellite RNA multimers were observed (KURODA et al. 1997). The junction between monomer units either was intact or contained deletions, predominantly of (+) 5′-terminal sequences, but also in some cases (+) 3′-terminal sequences (KURODA et al. 1997). Since circular forms of CMV satellite RNA were not detected in vivo (LINTHORST and KAPER 1984), it is assumed that the (+) satellite RNA multimers were produced by the CMV replicase via reinitiation of replication on the 3′ end of one (+) satellite RNA molecule before release of the nascent (+) satellite RNA molecule (KURODA et al. 1997), as originally proposed to explain the presence of multimers of turnip crinkle virus satellite RNA (CARPENTER et al. 1991).

In vitro-generated deletions of up to seven nucleotides from the (+) 3' terminus of a satellite RNA were repaired in planta by CMV, but not by TAV as the helper virus (BURGYÁN and GARCÍA-ARENAL 1998). This 3' repair was dependent on the presence of CMV RNAs 1 and 2 (BURGYáN and GARCÍA-ARENAL 1998), which encode the replicase proteins (NITTA et al. 1988; HAYES and BUCK 1990).

The effects of additional sequences at the termini of satellite RNA transcripts derived from biologically active cDNA clones were found to be different in different experiments. In several studies, such (+) RNA transcripts containing extra sequences at either the 3' terminus (COLLMER and KAPER 1986) or the 5' terminus (KURATH and PALUKAITIS 1987; MASUTA et al. 1987, 1988b) were infectious, although the infectivity was improved by removal of these additional sequences (MASUTA et al. 1987, 1988b). In one study, where extra sequences were present at both termini of (+) satellite RNA transcripts, either poor or no satellite RNA infection was observed (KURATH and PALUKAITIS 1987), while in another study, extra sequences at both termini of the (+) satellite RNA transcripts had no apparent effect on infectivity (TOUSCH et al. 1994). In transgenic plants expressing (+) sense monomer or greater than unit-length satellite RNA, the presence of extra sequences on both termini had no effect on the ability of such satellite RNAs to be replicated by the helper virus (BAULCOMBE et al. 1986; JACQUEMOND et al. 1988; TOUSCH et al. 1994).

In two studies, (–) satellite RNA transcripts (with extra nucleotides at both termini) were not infectious (COLLMER and KAPER 1986; KURATH and PALUKAITIS 1987), while in a third study such (–) RNA transcripts were found to initiate infection in only a low percentage of inoculated plants (TOUSCH et al. 1994). However, in those plants that were infected, the levels of ss and ds satellite RNA were similar to what was observed in plants inoculated with (+) satellite RNA transcripts (TOUSCH et al. 1994). In addition, in transgenic plants expressing (–) satellite RNAs containing extra sequences at both termini, CMV was able to replicate the satellite RNA, but the level of accumulated (+) satellite RNA was much lower (TOUSCH et al. 1994). Thus, it seems most likely that (–) satellite RNAs are poor templates for (+) satellite RNA synthesis in vivo, as was also observed in vitro (HAYES et al. 1992). This may be either because of the requirement for a single, additional guanosine residue at the 3' terminus of the (–) satellite RNA (WU and KAPER 1994), or because of a role for (+) satellite RNA in the biosynthesis of (+) satellite RNA. It is unlikely that the (–) satellite RNA is a poorer template than the (+) satellite RNA per se, since the ratio of (+):(–) satellite RNA that accumulates in infected plants is 2–3:1 (PIAZZOLLA et al. 1982b).

5.3 Replication In Vitro

A membrane-bound, RNA-dependent RNA polymerase (RdRp) was obtained from CMV (plus or minus) satellite RNA-infected tobacco, which produced various labeled forms of the CMV satellite RNA, although only of (+) polarity (YOUNG et al. 1987). This system also did not respond to added template. Template-dependent RdRp have been prepared from CMV-infected tobacco plants (HAYES and BUCK 1990; QUADT and JASPARS 1991). These RdRp were able to accept CMV satellite RNAs of (+) polarity, but not of (–) polarity (WU et al. 1991, 1993b; HAYES et al. 1992), unless there was an extra guanosine residue on the 3' end of the (–) satellite RNA (WU and KAPER 1994). This extra guanosine residue was first detected on the 3' end of ds satellite RNA as an unpaired residue (COLLMER and KAPER 1985). By contrast, when longer, nontemplate residues were present at both termini, (–) satellite RNA did not show template activity in vitro, and (+) satellite RNA was a template in vitro only for (–) RNA synthesis, with no ss (+) RNA synthesis (HAYES et al. 1992).

Although there is no doubt that the CMV RdRp replicates the satellite RNA in vivo, several studies show that there must be host factors that cause differential recognition of satellite RNAs as a template vs. the CMV RNAs: (a) In tomato, high temperatures affected the synthesis of (+) and (–) satellite RNA more than the synthesis of the CMV genomic RNAs (WHITE et al. 1995); (b) the Sny-strain of CMV did not support the replication of WL1-sat-RNA in squash, but did do so in tobacco, while there was no detectable debilitation of helper virus replication (GAL-ON et al. 1995); (c) co-infection of CMV containing a satellite RNA and tobacco etch virus (TEV) led to increases in the level of accumulation of CMV RNAs (i.e., synergy) but to suppression of the satellite RNA (PRUSS et al. 1997). Further studies of these systems could lead to a better understanding of the mechanism and specificity of satellite RNA replication.

6 Satellite Biocontrol

The observation that most satellite RNAs of CMV attenuate the symptoms induced by many strains of CMV (MOSSOP and FRANCKI 1979a; WATERWORTH et al. 1979; JACQUEMOND and LEROUX 1982) led to the idea of using satellite RNA as a biocontrol for the pathology of CMV, i.e., to induce tolerance to CMV. This has been done in two ways: either by transgenic expression of satellite RNA sequences, or by preinoculation of plants with "vaccines" of mild strains of CMV containing attenuating satellite RNAs. CMV-induced pathology was attenuated using either approach.

A CMV satellite that did not induce a pathological response, 117N-sat-RNA, was expressed transgenically either as 1.3 or 2.3 tandem units, and the satellite RNA was replicated and encapsidated by inoculated CMV (BAULCOMBE et al. 1986). The satellite RNA also attenuated CMV- or TAV-induced symptoms and reduced the accumulation of CMV, but not of TAV (HARRISON et al. 1987). Similar tolerance to CMV-induced disease and reduction in CMV accumulation was observed when a unit-length satellite RNA was expressed transgenically and when

CMV was introduced via an aphid vector (JACQUEMOND et al. 1988). Protection against CMV-induced disease has also been achieved in transgenic tomatoes expressing the satellite RNAs of CMV (SAITO et al. 1992; McGARVEY et al. 1994), but not against TAV-induced disease (McGARVEY et al. 1994). The extent of protection in tobacco varied with the dose of inoculum and the number of transgene copies (MASATA et al. 1994). In addition, two partial-length cDNA copies of S-sat-RNA showed only a 1- or 2-day delay in the appearance of symptoms, and differences in the extent of protection were observed between plants expressing a unit-length vs. a dimer-length cDNA copy of the S-sat-RNA (PEÑA et al. 1994). A higher level of resistance was observed when the coat protein gene of CMV as well as a satellite RNA of CMV were both expressed transgenically in tobacco than when either sequence alone was expressed (YIE et al. 1992).

The control of either CMV-induced or pathogenic, satellite-induced disease by viral vaccines, consisting of mild CMV strains containing an attenuating satellite RNA, has been demonstrated in several countries (JACQUEMOND 1982; YOSHIDA et al. 1985; TIEN et al. 1987; OHKI et al. 1989; WU et al. 1989; GALLITELLI et al. 1991; MONTASSER et al. 1991; TIEN and WU 1991; CRESCENZI et al. 1993b; SAYAMA et al. 1993). These include greenhouses as well as extensive field tests. Crop yields were increased by the viral vaccines (GALLITELLI et al. 1991; MONTASSER et al. 1991; TIEN and WU 1991; SAYAMA et al. 1993), and no adverse effects have been observed. In the case of mixed infections involving CMV and tobacco mosaic virus (TMV), potato virus X (PVX) or potato virus Y (PVY), where mixed infections can give rise to synergy, little or no differences in pathology were observed in mixed infected tomato (TIEN et al. 1987; SAYAMA et al. 1993) or pepper (TIEN et al. 1987) plants, although neither the level of CMV nor the presence of CMV satellite RNA was ascertained in such plants. Since the potyviruses TEV was shown to stimulate the accumulation of CMV and inhibit the accumulation of satellite RNA in tobacco (PRUSS et al. 1997), it might have been expected that the potyvirus PVY would show the same effect on the vaccine CMV and its satellite RNA; however, it is not clear whether the same effects occur in either tomato or pepper as in tobacco. If they occur in other host plants common to the above-mentioned virus combinations, then the application of the vaccines containing CMV and satellite RNAs could be potentially hazardous.

Since some satellite RNAs appear to be poorly supported at elevated temperatures (KAPER et al. 1995; WHITE et al. 1995), the strategy of using either CMV vaccines containing satellite RNAs or transgenic plants expressing satellite RNAs might seem problematic in countries with high day and night temperatures. However, exposure to high day temperatures does not seem to be sufficient to inhibit satellite RNA-mediated protection (GALLITELLI et al. 1991; CRESCENZI et al. 1993b).

Another potential hazard of using satellite RNAs in vaccines that has been identified is the possibility of a non-pathogenic satellite RNA mutating to a pathogenic form (PALUKAITIS 1991). Such an effect has been demonstrated in the greenhouse, during linear passaging of CMV containing a non-pathogenic satellite RNA, where a single nucleotide change in the satellite RNA rendered the satellite

RNA nonpathogenic (PALUKAITIS and ROOSSINCK 1996). However, no changes in pathogenicity were observed in the field after 2 years, although the nucleotide sequence of the protecting satellite RNA did change at two positions (GALLITELLI et al. 1997). For differences in pathogenicity to occur, more changes at specific positions would have to occur. To what extent such changes would be selected remains unknown, but it would presumably depend on the helper virus and the host species (see Sect. 3). In artificial mixtures of satellite RNA, some species were preferentially selected (SMITH et al. 1992; PALUKAITIS and ROOSSINCK 1995; ROOSSINCK and PALUKAITIS 1995), depending on the nature of the satellite RNA, as well as the helper virus. In natural mixtures, the nature of the dominating variant changed as a result of passage in different hosts (GARCIA-LUQUE et al. 1984; KAPER et al. 1988; KURATH and PALUKAITIS 1989a). Further analysis of naturally passaged vaccine strains will undoubtedly provide more insight into this issue.

7 Concluding Remarks

The satellite RNAs of CMV have provided a remarkable system for analyzing (a) the relationship between RNA sequence/structure and pathogenicity, (b) RNA sequence diversity, (c) the evolution of RNA molecules in controlled experiments and in field situations, (d) the restrictions of secondary structure on the evolution of an RNA molecule, (e) RNA replication in vivo and in vitro, and (f) biocontrol strategies. While much has been elucidated and delineated, there are still many questions that need to be addressed: What are the host factors that interact with the satellite RNAs to induce pathological reactions? How do minor changes in satellite RNA nucleotide sequence or structure affect these interactions, leading either to no pathology or to altered pathology? (For example, how do single nucleotide sequence changes affect interactions leading to either white or yellow chlorosis, or to chlorosis in tobacco vs. tomato?) Is pathogenesis really limited to only several *Nicotiana* or *Lycopersicon* species for a given satellite RNA, or does it also occur in other plant species among the some 1000 host species of CMV? How does temperature affect pathogenicity of the satellite RNAs and differentially affect the replication of CMV vs. satellite RNA? Why do satellite RNAs supported by TAV not reduce the accumulation of TAV RNAs? How does synergy increase the accumulation of CMV but also inhibit the accumulation of satellite RNA? What host, helper virus, or environmental factors result in selection of one satellite RNA species over another? Does this selection operate at the level of replication, encapsidation, or movement? What TAV elements encoded by RNAs 1 and 2 affect the movement of particular satellite RNAs? The answers to these questions and the underlying mechanisms will illuminate new regulatory processes in plants, as well as produce a more detailed picture of virus evolution and the nature of factors which influence RNA: protein interactions.

Acknowledgement. We wish to thank Dr. Aurora Fraile, Madrid, for the data in Fig. 1.

References

Alonso-Prados JL, Aranda MA, Malpica JM, García-Arenal F, Fraile A (1998) Satellite RNA of cucumber mosaic cucumovirus spreads epidemically in natural populations of its helper virus. Phytopathology 88:520 524

Aranda MA, Fraile A, García-Arenal F (1993) Genetic variability and evolution of the satellite RNA of cucumber mosaic virus during natural epidemics. J Virol 67:5896 5901

Aranda MA, Fraile A, Dopazo J, Malpica JM, García-Arenal F (1997) Contribution of mutation and RNA recombination to the evolution of a plant pathogenic RNA. J Mol Evol 44:81 88

Avila-Rincon MJ, Collmer CW, Kaper JM (1986a) In vitro translation of cucumoviral satellites. I. Purification and nucleotide sequence of cucumber mosaic virus-associated RNA 5 from cucumber mosaic virus strain S. Virology 152:446 454

Avila-Rincon MJ, Collmer CW, Kaper JM (1986b) In vitro translation of cucumoviral satellites. II. CARNA5 from cucumber mosaic virus strains S and SP6 transcripts of cloned (S) CARNA 5 cDNA produce electrophoretically co-migrating protein products. Virology 152:455 458

Baulcombe DC, Saunders GR, Bevan MW, Mayo MA, Harrison BD (1986) Expression of biologically active viral satellite RNA from the nuclear genome of transformed plants. Nature 321:446 449

Bernal JJ, García-Arenal F (1994a) Analysis of the satellite RNA of cucumber mosaic cucumovirus for high accumulation in squash. Virology 205:262 268

Bernal JJ, García-Arenal F (1994b) Complex interactions among several nucleotide positions determine phenotypes defective for long-distance transport in the satellite RNA of cucumber mosaic virus. Virology 200:148 153

Bernal JJ, García-Arenal F (1997) Analysis of the in vitro secondary structure of cucumber mosaic virus satellite RNA. RNA 3:1052 1067

Bruening G (1990) Replication of satellite RNA of tobacco ringspot virus. Semin Virol 1:127 134

Burgyán J, García-Arenal F (1998) Template-independent repair of the 3' end of cucumber mosaic virus satellite RNA controlled by the RNAs 1 + 2 of helper virus. J Virol (to be published)

Carpenter CD, Cascone PJ, Simon AE (1991) Formation of multimers of linear satellite RNAs. Virology 183:586 594

Collmer CW, Kaper JM (1985) Double-stranded RNAs of cucumber mosaic virus and its satellite contain an unpaired terminal guanosine: implications for replication. Virology 145:249 259

Collmer CW, Kaper JM (1986) Infectious RNA transcripts from cloned cDNAs of cucumber mosaic viral satellites. Biochem Biophys Res Commun 135:290 296

Collmer CW, Tousignant ME, Kaper JM (1983) Cucumber mosaic virus-associated RNA 5. X. The complete nucleotide sequence of a CARNA5 incapable of inducing tomato necrosis. Virology 127:230 234

Crescenzi A, Grieco F, Gallitelli D (1992) Nucleotide sequence of a satellite RNA of a strain of cucumber mosaic virus associated with a tomato fruit necrosis. Nucleic Acids Res 20:2886

Crescenzi A, Barbarossa L, Cillo F, Di Franco A, Vovlas N, Gallitelli D (1993a) Role of cucumber mosaic virus and its satellite RNA in the etiology of tomato fruit necrosis in Italy. Arch Virol 131:321 333

Crescenzi A, Barbarossa L, Gallitelli D, Martelli GP (1993b) Cucumber mosaic cucumovirus populations in Italy under natural epidemic conditions and after a satellite-mediated protection test. Plant Dis 77:28 33

Devic M, Jaegle M, Baulcombe D (1989) Symptom production on tobacco and tomato is determined by two distinct domains of the satellite RNA of cucumber mosaic virus (strain Y). J Gen Virol 70:2765 2774

Devic M, Jaegle M, Baulcombe D (1990) Cucumber mosaic virus satellite RNA (strain Y): analysis of sequences which affect systemic necrosis on tomato. J Gen Virol 71:1443 1449

Diaz-Ruiz JR, Kaper JM (1977) Cucumber mosaic virus-associated RNA 5. III. Little sequence homology between CARNA 5 and helper virus. Virology 80:204 213

Domingo E, Holland JJ (1997) RNA virus mutation and fitness for survival. Annu Rev Microbiol 51:151–178

Domingo E, Sabo D, Taniguchi T, Weissmann C (1978) Nucleotide sequence heterogeneity in an RNA phage population. Cell 13:735–744

Felden B, Florentz C, Giegé R, Westhof E (1994) Solution structure of the 3′ end of brome mosaic virus genomic RNAs. Conformational mimicry with canonical tRNAs. J Mol Biol 253:508–531

Fraile A, García-Arenal F (1991) Secondary structure as a constraint on the evolution of a plant viral satellite RNA. J Mol Biol 221:1065–1069

Fraile A, Moriones E, García-Arenal F (1990) Characterization of a satellite RNA associated with strain K8 of cucumber mosaic virus. Nucleic Acids Res 18:4593

Fraile A, Alonso-Prados JL, Aranda MA, Bernal JJ, Malpica JM, García-Arenal F(1997) Genetic exchange by recombination or reassortment is infrequent in natural populations of a tripartite RNA plant virus. J Virol 71:934–940

Gallitelli D, Franco AD, Vovlas C, Kaper JM (1988) Infezioni miste del virus del mosaico del centriolo (CMV) e di potyvirus in colture ortive di Puglia e Basilicata. Inform Fitopatol 38:57–64

Gallitelli D, Grieco F, Cillo F (1997) The potential of a beneficial satellite RNA of cucumber mosaic virus to acquire deleterious functions: nature versus greenhouses. In: Tepfer M, Balázs E (eds) Virus-resistant transgenic plants: potential ecological impact. Springer, Berlin Heidelberg New York, pp 100–106

Gallitelli D, Vovlas C, Martelli G, Montasser MS, Tousignant ME, Kaper JM (1991) Satellite-mediated protection of tomato against cucumber mosaic virus. II. Field test under natural epidemic conditions in southern Italy. Plant Dis 75:93–95

Gal-On A, Kaplan I, Palukaitis P (1995) Differential effects of satellite RNA on the accumulation of cucumber mosaic virus RNAs and their encoded protein in tobacco vs. zucchini squash with two strains of CMV helper virus. Virology 208:58–68

García-Arenal F, Zaitlin M, Palukaitis P (1987) Sequence analysis of six satellite RNAs of cucumber mosaic virus differing in their pathology. Virology 158:339–347

García-Luque I, Kaper JM, Diaz-Ruiz JR, Rubio-Huertos M (1984) Emergence and characterization of satellite RNAs associated with Spanish cucumber mosaic virus isolates. J Gen Virol 65:539–547

Gonsalves D, Provvidenti R, Edwards MC (1982) Tomato white leaf: the relation of an apparent satellite RNA and cucumber mosaic virus. Phytopathology 72:1533–1538

Gordon KHJ, Symons RH (1983) Satellite RNA of cucumber mosaic virus forms a secondary structure with partial 3′-terminal homology to genomal RNA. Nucleic Acids Res 11:947–960

Gould AR, Palukaitis P, Symons RH, Mossop DW (1978) Characterization of a satellite RNA associated with cucumber mosaic virus. Virology 85:443–455

Grieco F, Lanave C, Gallitelli D (1997) Evolutionary dynamics of cucumber mosaic virus satellite RNA during natural epidemics in Italy. Virology 229:166–174

Habili N, Kaper JM (1981) Cucumber mosaic virus-associated RNA 5. VII. Double-stranded form accumulation and disease attenuation in tobacco. Virology 112:250–261

Harrison BD, Mayo MA, Baulcombe DC (1987) Virus resistance in transgenic plants that express cucumber mosaic virus satellite RNA. Nature 328:799–802

Hayes RJ, Buck KW (1990) Complete replication of a eukaryotic virus RNA in vitro by a purified RNA-dependent RNA polymerase. Cell 63:763–368

Hayes RJ, Tousch D, Jacquemond M, Pereira VC, Buck KW, Tepfer M (1992) Complete replication of a satellite RNA in vitro by a purified RNA-dependent RNA polymerase. J Gen Virol 73:1597–1600

Hidaka S, Hanada K (1994) Structural features unique to a new 405-nucleotide satellite RNA of cucumber mosaic virus inducing tomato necrosis. Virology 200:806–808

Hidaka S, Ishikawa K, Takanami,Y, Kubo S, Miura K (1984) Complete nucleotide sequence of RNA5 from cucumber mosaic virus (strain Y). FEBS Lett 174:38–42

Hidaka S, Hanada K, Ishikawa K, Miura K (1988) Complete nucleotide sequence of two new satellite RNAs associated with cucumber mosaic virus. Virology 164:326–333

Hidaka S, Hanada K, Ishikawa K (1990) In vitro messenger properties of a satellite RNA of cucumber mosaic virus. J Gen Virol 71:439–442

Jacquemond M (1982) Phénomènes d'interférences entre les deux types d'ARN satellite du virus de la mosa du concombre. Protection des tomates vis-à-vis de la nécrose létale. C R Acad Sci Paris 294:991–994

Jacquemond M, Leroux J-P (1982) L'ARN satellite du virus de la mosa du concombre II. Etude de la relation virus-ARN satellite chez divers hôtes. Agronomie 2:55–62

Jacquemond M, Lot H (1981) L'ARNA satellite du virus de la la mosa du concombre I. Comparison de l'aptitude à induire la nécrose de la tomate d'ARN satellite isolés de plusieurs souches du virus. Agronomie 1:927–932

Jacquemond M, Lot H (1982) L'ARN satellite du virus de la mosa du concombre. III. La propriété de survie in vivo. Agronomie 2:533–538

Jacquemond M, Amselem J, Tepfer M (1988) A gene coding for a monomeric form of cucumber mosaic virus satellite RNA confers tolerance to CMV. Mol Plant Microbe Interact 1:311–316

Jaegle M, Devic M, Longstaff M, Baulcombe D (1990) Cucumber mosaic virus satellite RNA (Y strain): analysis of sequences which affect yellow mosaic symptoms on tobacco. J Gen Virol 71:1905–1912

Jorda C, Alfaro A, Aranda MA, Moriones E, García-Arenal F (1992) Epidemic of cucumber mosaic virus plus satellite RNA in tomatoes in eastern Spain. Plant Dis 76:363–366

Jukes TH, Cantor CR (1969) Evolution of protein molecules. In: Munro HN (ed) Mammalian protein metabolism. Academic, New York, pp 21–132

Kaper JM (1982) Rapid synthesis of double-stranded cucumber mosaic virus-associated RNA 5: mechanism controlling viral pathogenesis? Biochem Biophys Res Commun 105:1014–1022

Kaper JM, Tousignant ME (1977) Cucumber mosaic virus-associated RNA 5. I. Role of host plant and helper strain in determining amount of associated RNA 5 with virions. Virology 80:186–195

Kaper JM, Tousignant ME (1978) Cucumber mosaic virus-associated RNA5. V. Extensive nucleotide sequence homology among CARNA 5 preparations of different CMV strains. Virology 85:323–327

Kaper JM, Waterworth HE (1977) Cucumber mosaic virus-associated RNA 5: causal agent for tomato necrosis. Science 196:429–431

Kaper JM, Tousignant ME, Lot H (1976) A low molecular weight replicating RNA associated with a divided genome plant virus: defective or satellite RNA? Biochem Biophys Res Commun 72:1237–1243

Kaper JM, Tousignant ME, Thompson S (1981) Cucumber mosaic virus-associated RNA 5. VIII. Identification and partial characterization of a CARNA 5 incapable of inducing tomato necrosis. Virology 114:526–533

Kaper JM, Duriat AS, Tousignant ME (1986) The 368-nucleotide satellite of cucumber mosaic virus strain Y from Japan does not cause lethal necrosis in tomato. J Gen Virol 67:2241–2246

Kaper JM, Tousignant ME, Steen MT (1988) Cucumber mosaic virus-associated RNA 5. XI. Comparison of 14 CARNA 5 variants relates ability to induce tomato necrosis to a conserved nucleotide sequence. Virology 163:284–292

Kaper JM, Tousignant ME, Geletka LM (1990) Cucumber-mosaic-virus-associated RNA-5. XII. Symptom-modulating effect is codetermined by the helper virus satellite replication support function. Res Virol 141:487–503

Kaper JM, Geletka LM, Wu GS, Tousignant ME (1995) Effect of temperature on cucumber mosaic virus satellite-induced lethal tomato necrosis is helper virus strain dependent. Arch Virol 140:65–74

Kearney CM, Zitter TA, Gonsalves D (1990) A field survey for serogroups and satellite RNA of cucumber mosaic virus. Phytopathology 80:1238–1243

Kosaka Y, Hanada K, Fukunishi T, Tochihara H (1989) Cucumber mosaic virus isolate causing tomato necrotic disease in Kyoto Prefecture. Ann Phytopathol Soc Jpn 55:229–232

Kurath G, Palukaitis P (1987) Biological activity of T7 transcripts of a prototype clone and a sequence variant clone of a satellite RNA of cucumber mosaic virus. Virology 159:199–208

Kurath G, Palukaitis P (1989a) RNA sequence heterogeneity in natural populations of three satellite RNAs of cucumber mosaic virus. Virology 173:231–240

Kurath G, Palukaitis P (1989b) Satellite RNAs of cucumber mosaic virus: recombinants constructed in vitro reveal independent functional domains for chlorosis and necrosis in tomato. Mol Plant Microbe Interact 2:91–96

Kurath G, Palukaitis P (1990) Serial passage of infectious transcripts of a cucumber mosaic virus satellite RNA clone results in sequence heterogeneity. Virology 176:8–15

Kuroda T, Natsuaki T, Wang W-Q, Okuda S (1997) Formation of multimers of cucumber mosaic virus satellite RNA. J Gen Virol 78:941–946

Kuwata S, Masuta C, Takanami Y (1991) Reciprocal phenotype alterations between two satellite RNAs of cucumber mosaic virus. J Gen Virol 72:2385–2389

Lee HS, Kummert J (1985) Induction of tomato necrosis by cucumoviruses, as related to specific interactions between genomic and satellite RNAs. Parasitica 41:45–55

Li WH, Gojobori T, Nei M (1981) Pseudogenes as a paradigm of neutral evolution. Nature 292:237–239

Linthorst HJM, Kaper JM (1984) Replication of peanut stunt virus and its associated RNA 5 in cowpea protoplasts. Virology 139:317–329

Linthorst HJM, Kaper JM (1985) Cucumovirus satellite RNAs cannot replicate autonomously in cowpea protoplasts. J Gen Virol 66:1839–1842

Marrou J, Duteil M (1974) La necrose de la tomate. I. Reproduction des symptomes de la maladie par inoculation mecanique de plusieurs souches du virus de la mosa du concombre (CMV). Ann Phytopathol 6:155–171

Marrou J, Duteil M, Lot H, Clerjeau H (1973) La nécrose de la tomate: une grave virose des tomates cultivées en plein champ. Pepin Hortic Maraich 137:37–41

Masuta C, Takanami Y (1989) Determination of sequence and structural requirements for pathogenicity of a cucumber mosaic virus satellite RNA (Y-satellite RNA). Plant Cell 1:1165–1173

Masuta C, Kuwata S, Takanami Y (1987) In vitro synthesis of infectious RNAs from cDNAs of cucumber mosaic virus satellite RNA (strain Y) after removal of non-viral bases with ribonuclease H. Nucleic Acids Res 15:10048

Masuta C, Kuwata S, Takanami Y (1988a) Disease modulation on several plants by cucumber mosaic virus satellite RNA (Y strain). Ann Phytopathol Soc Jpn 54:332–336

Masuta C, Kuwata S, Takanami Y (1988b) Effects of extra 5' non-viral bases on the infectivity of transcripts from a cDNA clone of satellite RNA (strain Y) of cucumber mosaic virus. J Biochem 104:841–846

Masuta C, Komari T, Takanami Y (1989) Expression of cucumber mosaic virus satellite RNA from cDNA copies in transgenic plants. Ann Phytopathol Soc Jpn 55:49–55

Masuta C, Kuwata S, Matzuzaki T, Takanami Y, Koiwai A (1992) A plant virus satellite RNA exhibits a significant sequence complementarity to a chloroplast tRNA. Nucleic Acids Res 20:2885

Masuta C, Suzuki M, Matsuzaki T, Honda I, Kuwata S, Takanami Y, Koiwai A (1993a) Bright yellow chlorosis by cucumber mosaic virus Y satellite RNA is specifically induced without severe chloroplast damage. Phys Mol Plant Pathol 42:267–278

Masuta C, Suzuki M, Kuwata S, Takanami Y, Koiwai A (1993b) Yellow mosaic symptoms induced by Y satellite RNA of cucumber mosaic virus is regulated by a single incompletely dominant gene in wild Nicotiana species. Phytopathology 83:411–413

Masuta C, Hayashi Y, Suzuki M, Kuwata S, Takanami Y, Koiwai A (1994) Protective effect of a satellite RNA expressed in transgenic plants on disease incidence after inoculation of cucumber mosaic virus. Ann Phytopathol Soc Jpn 60:228–232

McGarvey PB, Kaper JM, Avila-Rincón MJ, Peña L, Diaz-Ruiz JR (1990) Transformed tomato plants express a satellite RNA of cucumber mosaic virus and produce lethal necrosis upon infection with viral RNA. Biochem Biophys Res Commun 170:548–555

McGarvey PB, Montasser MS, Kaper JM (1994) Transgenic tomato plants expressing satellite RNA are tolerant to some strains of cucumber mosaic virus. J Am Soc Hort Sci 119:642–647

Montasser MS, Tousignant ME, Kaper JM (1991) Satellite-mediated protection of tomato against cucumber mosaic virus. I. Greenhouse experiments and simulated epidemic conditions in the field. Plant Dis 75:86–92

Moriones E, Fraile A, Garcia-Arenal F (1991) Host-associated selection of sequence variants from a satellite RNA of cucumber mosaic virus. Virology 184:465–468

Moriones E, Diaz I, Rodriguez-Cerezo E, Fraile A, Garcia-Arenal F (1992) Differential interactions among strains of tomato aspermy virus and satellite RNAs of cucumber mosaic virus. Virology 186:475–480

Moriones E, Diaz I, Fernandez-Cuartero B, Fraile A, Burgyan J, Garcia-Arenal F (1994) Mapping helper functions for cucumber mosaic virus satellite RNA with pseudorecombinants derived from cucumber mosaic and tomato aspermy viruses. Virology 205:574–577

Mossop DW, Francki RIB (1978) Survival of a satellite RNA in vivo and its dependence on cucumber mosaic virus for replication. Virology 86:562–566

Mossop DW, Francki RIB (1979a) Comparative studies on two satellite RNAs of cucumber mosaic virus. Virology 95:395–404

Mossop DW, Francki RIB (1979b) The stability of satellite viral RNAs in vivo and in vitro. Virology 94:243–253

Nitta N, Takanami Y, Kuwata S, Kubo S (1988) Inoculation with RNAs 1 and 2 of cucumber mosaic virus induces viral RNA replicase activity in tobacco mesophyll protoplasts. J Gen Virol 69:2695–2700

Ohki ST, Tanaka H, Inouye T (1989) Cucumber mosaic virus satellite RNA transmissible to plants infected with a different isolate of CMV. Ann Phytopathol Soc Jpn 55:69–71

Owens RA, Steger G, Hu Y, Fels A, Hammonds RW, Riesner D (1996) RNA structural features responsible for potato spindle tuber viroid pathogenicity. Virology 222:144–258

Palukaitis P (1988) Pathogenicity regulation by satellite RNAs of cucumber mosaic virus: minor nucleotide sequence changes alter host responses. Mol Plant Microbe Interact 1:175–181

Palukaitis P (1991) Virus-mediated genetic transfer in plants. In: Levin M, Strauss HS (eds) Risk assessment in genetic engineering: environmental release of organisms. McGraw-Hill, New York, pp 140–162

Palukaitis P, Roossinck MJ (1995) Variation in the hypervariable region of cucumber mosaic virus satellite RNAs is affected by the helper virus and the initial sequence context. Virology 206:765–768

Palukaitis P, Roossinck MJ (1996) Spontaneous change of a benign satellite RNA of cucumber mosaic virus to a pathogenic variant. Nat Biotech 14:1264–1268

Peña L, Trad J, Díaz-Ruiz JR, McGarvey PB, Kaper JM (1994) Cucumber mosaic virus protection in transgenic tobacco plants expressing monomeric, dimeric or partial sequences of a benign satellite RNA. Plant Sci 100:71–82

Piazzolla P, Gallitelli D, Savino V (1982a) Appearance of satellite RNA (CARNA 5) in six cucumber mosaic virus isolates from the open field. Phytopathol Medit 21:32–34

Piazzolla P, Tousignant ME, Kaper JM (1982b) Cucumber mosaic virus-associated RNA 5. IX. The overtaking of viral RNA synthesis by CARNA 5 and dsCARNA 5 in tobacco. Virology 122:147–157

Pleij CWA, Abrahams JP, Van Belkum A, Rietveld K, Bosch L (1987) The spatial folding of the 3'-noncoding region of aminoacylatable plant viral RNAs. In: Brinton RA, Rueckert RR (eds) Positive-strand RNA viruses. Liss, New York, pp 199–316

Pruss G, Ge X, Shi XM, Carrington JC, Vance VB (1997) Plant viral synergism: the potyviral genome encodes a broad-range pathogenicity enhancer that transactivates replication of heterologous viruses. Plant Cell 9:859–868

Quadt R, Jaspars EMJ (1991) Characterization of cucumber mosaic virus RNA-dependent RNA polymerase. FEBS Lett 279:273–276

Richards KE, Jonard G, Jacquemond M, Lot H (1978) Nucleotide sequence of cucumber mosaic virus-associated RNA 5. Virology 89:395–408

Rodríguez-Alvarado G, Roossinck MJ (1997) Structural analysis of a necrogenic strain of cucumber mosaic cucumovirus satellite RNA in planta. Virology 236:155–166

Rodríguez-Cerezo E, García-Arenal F (1989) Genetic heterogeneity of the RNA genome population of the plant virus U5-TMV. Virology 170:418–423

Roossinck MJ, Palukaitis P (1995) Genetic analysis of helper virus-specific selective amplification of cucumber mosaic virus satellite RNAs. J Mol Evol 40:25–29

Roossinck MJ, Sleat D, Palukaitis P (1992) Satellite RNAs of plant viruses: structure and biological effects. Microbiol Rev 56:265–279

Roossinck MJ, Kaplan I, Palukaitis P (1997) Support of a cucumber mosaic virus satellite RNA maps to a single amino acid proximal to the helicase domain of the helper virus. J Virol 71:608–612

Saito Y, Komari T, Masuta C, Hayashi Y, Kumashiro T, Takanami Y (1992) Cucumber mosaic virus-tolerant transgenic tomato plants expressing a satellite RNA. Theor Appl Genet 83:679–683

Sayama H, Sato T, Kominato M, Natsuaki T, Kaper JM (1993) Field testing of a satellite-containing attenuated strain of cucumber mosaic virus for tomato protection in Japan. Phytopathology 83:405–410

Sleat DE (1990) Nucleotide sequence of a new satellite RNA of cucumber mosaic virus. Nucleic Acids Res 18:3416

Sleat DE, Palukaitis P (1990a) Induction of tobacco chlorosis by certain cucumber mosaic virus satellite RNAs is specific to subgroup II helper strains. Virology 176:292–295

Sleat DE, Palukaitis P (1990b) Site-directed mutagenesis of a plant viral satellite RNA changes its phenotype from ameliorative to necrogenic. Proc Natl Acad Sci U S A 87:2946–2950

Sleat DE, Palukaitis P (1992) A single nucleotide change within a plant virus satellite RNA alters the host specificity of disease induction. Plant J 2:43–49

Sleat DE, Zhang L, Palukaitis P (1994) Mapping determinants within cucumber mosaic virus and its satellite RNAs for the induction of necrosis in tomato plants. Mol Plant Microbe Interact 7:189–195

Smith CR, Tousignant ME, Geletka LM, Kaper JM (1992) Competition between cucumber mosaic virus satellite RNAs in tomato seedlings and protoplasts: a model for a satellite-mediated control of tomato necrosis. Plant Dis 76:1270–1274

Takahashi H, Ehara Y (1993) Severe chlorotic spot symptoms in cucumber mosaic virus strain Y-infected tobaccos are induced by a combination of the virus coat protein and two recessive genes. Mol Plant Microbe Interact 6:182–189

Takanami Y (1981) A striking change in symptoms on cucumber mosaic virus-infected tobacco plants induced by a satellite RNA. Virology 109:120–126

Taliansky ME, Ryadov EV, Robinson DJ, Palukaitis P (1988) Tomato cell death mediated by complementary plant viral satellite RNA sequences. Mol Plant Microbe Internet 11: (in press)

Tien P, Wu G (1991) Satellite RNA for the biocontrol of plant disease. Adv Virus Res 39:321–339

Tien P, Xhang X, Qiu B, Qin B, Wu G (1987) Satellite RNA for the control of plant diseases caused by cucumber mosaic virus. Ann Appl Biol 111:143–152

Tousch D, Jacquemond M, Tepfer M (1994) Replication of cucumber mosaic virus satellite RNA from negative-sense transcripts produced either in vitro or in transgenic plants. J Gen Virol 75:1009–1014

Tousignant ME, Kaper JM (1993) Cucumber mosaic virus-associated RNA 5. XIII. Opposite necrogenicities in tomato of variants with large 5′ half insertion/deletion regions. Res Virol 144:349–360

Waterworth HE, Kaper JM, Tousignant ME (1979) CARNA 5, the small cucumber mosaic virus-dependent replicating RNA, regulates disease expression. Science 204:845–847

White JL, Kaper JM (1987) Absence of lethal stem necrosis in select Lycopersicon spp. infected by cucumber mosaic virus strain D and its necrogenic satellite CARNA 5. Phytopathology 77:808–811

White JL, Tousignant ME, Geletka LM, Kaper JM (1995) The replication of a necrogenic cucumber mosaic virus satellite is temperature-sensitive in tomato. Arch Virol 140:53–63

Wu G, Kaper JM (1992) Widely separated sequence elements within cucumber mosaic virus satellites contribute to their ability to induce lethal tomato necrosis. J Gen Virol 73:2805–2812

Wu G, Kaper JM (1994) Requirement of 3′-terminal guanosine in (−)-stranded RNA for in vitro replication of cucumber mosaic virus satellite RNA by viral RNA-dependent RNA polymerase. J Mol Biol 238:655–657

Wu G, Kang L, Tien P (1989) The effect of satellite RNA on cross-protection among cucumber mosaic virus strains. Ann Appl Biol 114:489–496

Wu G, Kaper JM, Jaspars EMJ (1991) Replication of cucumber mosaic virus satellite RNA in vitro by an RNA-dependent RNA polymerase from virus-infected tobacco. FEBS Lett 292:213–216

Wu G, Kaper JM, Tousignant ME, Masuta C, Kuwata S, Takanami Y, Peña L, Diaz-Ruiz JR (1993a) Tomato necrosis and the 369 nucleotide Y satellite of cucumber mosaic virus: factors affecting satellite biological expression. J Gen Virol 74:161–168

Wu G, Kaper JM, Kung SD (1993b) Replication of satellite RNA in vitro by homologous and heterologous cucumoviral RNA-dependent RNA polymerases. Biochemie 75:749–755

Yie Y, Zhao F, Zhao SZ, Liu YZ, Liu YL, Tien P (1992) High resistance to cucumber mosaic virus conferred by satellite RNA and coat protein in transgenic commercial tobacco cultivar G-140. Mol Plant Microbe Interact 5:460–465

Yoshida K, Gotto T, Iizuka N (1985) Attenuated isolates of cucumber mosaic virus produced by satellite RNA and cross-protection between attenuated isolates and virulent ones. Ann Phytopathol Soc Jpn 51:238–242

Young ND, Palukaitis P, Zaitlin M (1987) Characterization of multimeric forms of cucumber mosaic virus satellite RNA. Proceedings of the UCLA symposium. In: Arntzen CJ, Ryan C (eds.) Molecular strategies for crop protection. Liss, New York, pp 243–252

Zhang L, Kim CH, Palukaitis P (1994) The chlorosis-induction domain of the satellite RNA of cucumber mosaic virus: identifying sequences that affect accumulation and the degreee of chlorosis. Mol Plant Microbe Interact 7:208–213

Large Satellite RNA: Molecular Parasitism or Molecular Symbiosis

M.A. Mayo,[1] M.E. Taliansky,[1] and C. Fritsch[2]

1 Introduction

Just as relationships among cellular organisms can involve several species and various levels of dependency, so it is for subcellular agents. For viruses, the dependent member of one such relationship has been termed "satellite" based on the "satellite virus" discovered by KASSANIS (1962) in some cultures of tobacco necrosis virus. Since then, satellites have been found in association with diverse viruses that infect most types of hosts (MAYO et al. 1995). The classic view of virus satellites, formed during the several years in which the only example was of satellite tobacco necrosis virus, is that they are parasitic on a virus and cause some penalty on the biological success of the helper. The satellites described in other chapters of this volume are very largely of this type. Indeed, this effect was the basis for attempts to use satellites to control diseases induced by the helper, cucumber mosaic virus. These were either by prophylactic inoculation with satellite-bearing virus (e.g., TIEN PO and CHANG 1983) or by transformation

[1] Scottish Crop Research Institute, Invergowrie, Dundee DD2 5DA, UK
[2] Institut de Biologie Moléculaire des Plantes du CNRS, 12 rue du Général Zimmer, 67084 Strasbourg Cedex, France
MAM and MET are supported by funding from the Scottish Office Agriculture Environment and Fisheries Department

of plants to be protected with satellite-specific cDNA (e.g., BAULCOMBE et al. 1986; HARRISON et al. 1987). But satellites differ in the nature of their dependency and in the degree to which the helper virus is 'parasitized'. Some are relatively benign and seem to be well adapted to an unassuming harmonious relationship with the helper, neither helping nor hindering its multiplication. This chapter focuses on such satellites, namely the "large mRNA satellites" (MAYO et al. 1995) found associated with nepoviruses and the satellite RNA of the potexvirus bamboo mosaic virus (BaMV).

It is a relatively small evolutionary step for a satellite to change from having no deleterious effects for the helper virus, to having beneficial effects. And it is a small further step for the helper virus to come to depend on the satellite-coded function for its biological success. Such RNAs cannot be described as inessential for the helper virus, and thus cannot be considered as satellites. Instead, they are grouped together as "satellite-like RNAs". An example of this relationship is the association of the umbravirus, groundnut rosette virus (GRV), with a satellite-like RNA that is essential for the transmission of GRV by aphids. However, a structurally very similar RNA is associated with pea enation mosaic virus (PEMV), but it is not essential for the helper and is thus a "true satellite". This is discussed in more detail in Sect. 3.1. Similarly, beet necrotic yellow vein virus (BNYVV) in nature has a four- or five-part genome but in laboratory culture can, and does, lose RNA components, sometimes becoming a bipartite genome virus (see Sect. 3.2).

There are a variety of associations between viruses in nature that resemble this sort of benign satellite relationship. Transmission of one virus can depend on a gene product of another virus that may be closely, distantly, or not at all related to it. For example, potato virus C depends on a second potyvirus, potato virus Y (KASSANIS and GOVIER 1971); the sequivirus parsnip yellow fleck virus, depends on another member of the family *Sequiviridae*, the waikavirus anthriscus yellows virus; and the dsDNA genome badnavirus, rice tungro bacilliform virus, depends on the ssRNA genome waikavirus rice tungro spherical virus (HARRISON and MURANT 1984).

Mutually dependent viruses are the next step in this imagined developmental sequence. The most vivid example is that of PEMV (reviewed by DEMLER et al. 1996b). The disease it induces results from infection by two RNA components, each of which encodes a replicase. Each component can replicate in the laboratory independently of the other, and current taxonomic thinking is that each should be classified in a separate genus (D'ARCY and MAYO 1997). But the disease results only from a combination of what have been proposed to be species of viruses from very different genera. These systems are discussed in comparison to 'true satellites', and also compared with the well-recognised symbiotic relationship between the genome components of a number of bipartite genome viruses.

2 True Satellites

2.1 mRNA Satellites of Nepoviruses

Currently, it is mainly nepoviruses that are associated with the satellite RNAs that are mRNAs (Table 1). Indeed, the third satellite system to be discovered, after those of satellite tobacco necrosis virus (KASSANIS 1962) and satellite tobacco ringspot virus (SCHNEIDER 1969), was that of satellite RNA of the nepovirus, tomato black ring virus (TBRV; MURANT et al. 1973).

Large satellite RNAs associated with nepoviruses (Table 1) are generally present as a minority component of the RNA present in virus particles, and such satellites do not greatly affect the pathogenicity of the helper virus. For example, the presence of the satellite RNA of arabis mosaic virus (ArMV) in virus cultures exacerbated symptoms in three hosts and ameliorated symptoms in ten species but had no effect in 29 species (LIU et al. 1991b). However, in none of these hosts did the satellite have a striking effect on virus accumulation. Even though satellite RNA of grapevine fanleaf virus (GFLV) is atypical, in that it can comprise 32% of the RNA in virus particles, it does not seem to have much effect on virus accumulation in infected leaves of *Chenopodium quinoa*.

Satellite RNAs in this category are linear molecules of 0.8–1.5kb (Table 1) and, like the genome RNAs of their helper viruses, have a 3'-terminal poly(A) and a 5'-linked VPg (FRITSCH et al. 1993). Sequences are known for eight of the ten nepovirus satellite RNAs (Table 1), but these reveal little homology between the different satellite RNAs, except for a U/AUGAAAA sequence at the 5' end, which

Table 1. Satellite and satellite-like RNA relevant to this chapter

	Size (nt)	ORF	Reference
Satellite RNA			
Arabis mosaic virus large satellite RNA	1104	39K	1
Bamboo mosaic virus satellite RNA	836	20K	2
Chicory yellow mottle virus large satellite RNA	1145	40K	3
Grapevine Bulgarian latent virus satellite RNA	ca. 1500	?	4
Grapevine fanleaf virus satellite RNA	1114	37K	5
Myrobalan latent ringspot virus satellite RNA	ca. 1400	45K	6
Pea enation mosaic virus satellite RNA	717		7
Strawberry latent ringspot virus satellite RNA	1118	36.5K	8
Tomato black ring virus satellite RNAs	1372–1376	48K	9
Satellite-like RNA			
Groundnut rosette virus satellite-like RNA	900	--	10
Beet necrotic yellow vein virus RNA-3	1774	25K	11
Beet necrotic yellow vein virus RNA-4	1467	31K	11
Beet necrotic yellow vein virus RNA-5	1342–1349	26K	12

References: *1* LIU et al. 1990, *2* LIN and HSU 1994, *3* RUBINO et al. 1990, *4* GALLITELLI et al. 1983, *5* FUCHS et al. 1989, *6* FRITSCH et al. 1984, *7* DEMLER et al. 1994, *8* KREIAH et al. 1993, *9* HEMMER et al. 1987, *10* BLOK et al. 1994, *11* BOUZOUBAA et al. 1985, *12* KIGUCHI et al. 1996.

is also present in the nepovirus genomic RNAs (Fuchs et al. 1989). All these satellite RNAs encode a nonstructural protein that can be synthesized by in vitro translation systems (Liu et al. 1991a; Rubino et al. 1990; Fritsch et al. 1978, 1980; Pinck et al. 1988). The protein has also been detected in vivo, in both infected tobacco protoplasts (TBRV: Fritsch et al. 1978) and infected *C. quinoa* leaves (GFLV; Moser et al. 1992).

Comparisons among the amino acid sequences of the proteins encoded by nepovirus satellite RNAs did not reveal shared sequence features, except that all proteins contained numerous phosphorylation motifs of the type S/TxxD/E and S/TxR/K (Leader and Katan 1988). However, there were some sequence matches between parts of the proteins of different subsets of the proteins of TBRV, ArMV, GFLV, SLRV, and CyMV (Fritsch et al. 1993). Another feature was that all these proteins are relatively basic, with a net positive charge of between 24 and 34 (Fritsch et al. 1993). This is the degree of basicity found in histones, and therefore tentatively suggests that these proteins have a high affinity for RNA. More precise analyses of charge distribution revealed three domains: a central region containing basic and acidic residues surrounded by two strongly basic N- and C-terminal regions (Fritsch et al. 1993). Comparisons among TBRV, GFLV, and ArMV satellite proteins using the vectorial representation algorithm resulted in a visualization of two average slopes which may delimit two functional domains (Fuchs et al. 1989). The first domain that comprises 150 amino acids is possibly involved in nucleic acid binding. Northwestern blot experiments, using as a probe TBRV satellite RNA protein (strain C) produced in *Escherichia coli*, have confirmed that the protein binds strongly to nucleic acids. Unexpectedly, however, this affinity was retained by protein that lacked N-terminal sequences, which suggests that regions other than this highly basic domain contribute to the high affinity (C. Oncino, personal communication).

Mutagenesis of full-length cDNA clones of the satellite RNAs of ArMV, GFLV and TBRV has shown that the encoded nonstructural proteins are required for the accumulation of satellite RNA in infected protoplasts. Thus, the proteins are involved in the replication of the satellite RNAs (Hans et al. 1993; Hemmer et al. 1993; Liu and Cooper 1993).

Like other satellites, large nepovirus satellite RNAs are relatively specific to their helper viruses. Satellite RNAs S and L of TBRV, which share 90% sequence identity, are able to replicate with the same helper virus (Oncino et al. 1995). Also GFLV satellite RNA, which is similar in sequence to ArMV satellite RNA, can use ArMV as a helper (Hans et al. 1993). In contrast, TBRV isolates of serotype S (Scottish) are unable to support the multiplication of satellite RNAs from TBRV isolates of serotype G (German), although the different satellite RNAs were about 60% identical. Moreover, the satellites were not interchangeable even though in their 5'- and 3'-noncoding regions the satellites are more similar than they are to the corresponding regions of the helper virus genomic RNAs (Oncino et al. 1995). The addition of satellite RNAs to pseudorecombinant isolates containing TBRV-S and TBRV-G RNAs showed that satellite RNA accumulation is correlated with the presence of genomic RNA-1 (which encodes the replication functions). This suggested that the interaction between satellite RNA and the helper virus replicase governs the specificity (MURANT and MAYO 1982).

More insights into the function of the satellite RNA-encoded proteins have been obtained by ONCINO et al. (1995) in experiments with TBRV satellite RNA. A deletion mutant that did not encode P48 did not multiply, even in cells co-infected with wild-type satellite RNA. This suggests that the synthesized wild-type P48 interacts only in *cis* with the RNA molecule from which it was synthesized. Analyses of the replication of chimeric satellite RNA constructed by exchanging different regions between satellite RNAs C and L (Fig. 1) have shown that, whereas exchanges of approximately the 380 5'-terminal nucleotides, or of approximately the 190 3'-terminal nucleotides (Fig. 1) did not destroy the ability of the satellite RNA to replicate with the helper virus, all exchanges done between these two regions produced an RNA that was unable to replicate with either of the helpers. Exchange of only the 5'-noncoding region also produced nonfunctional RNA. These results show that although 5'- and 3'-noncoding regions play an important role in the replication of the satellite RNA, the central region of the satellite RNA-encoded protein contains determinants which may react specifically with the viral replicase. The fact that about 350 nucleotides within the coding region had to be exchanged concomitantly with the 5'-noncoding region for the RNA to be replicated, suggests that the N-terminal region of the satellite RNA protein is involved in the specific recognition of the 5' end of the RNA during the synthesis of the (+) strand from the (−) strand, and/or (more likely) that the exchange of the first 380 nucleotides has conserved, by chance, a conformation recognizable by the viral

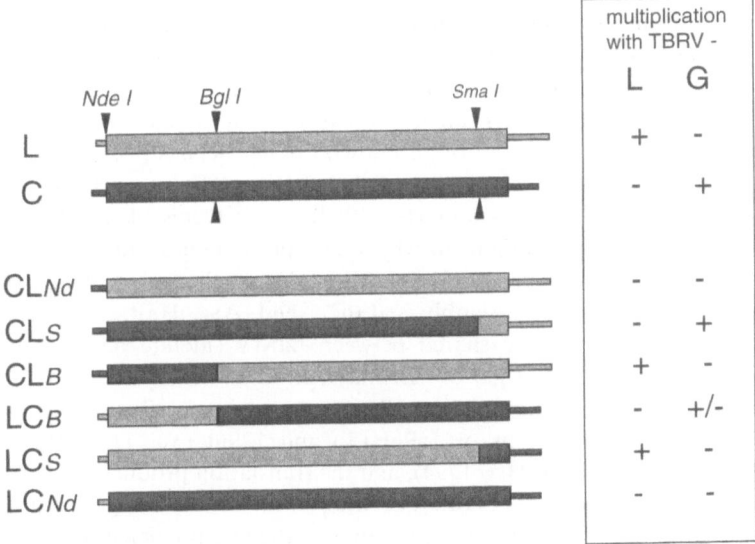

Fig. 1. Chimeric satellite RNAs. The parental satellite sequences were those of TBRV-C (*dark shading*) and TBRV-L (*light shading*)

replicase. It has also been shown that the 5'- and 3'-noncoding regions are important in the replication of the satellite RNAs of GFLV (Hans et al. 1993) and ArMV (Liu et al. 1991a).

A hypothetical model of replication of the satellite RNA has been proposed by Oncino et al. (1995) in which the P48 acts as a cofactor to modify the configuration of the satellite RNA, so as to make it recognizable by the viral replicase. Thus, in this model, the interaction between the satellite RNA-encoded protein and the viral replicase involves specific signals contained in each of the partners.

2.2 Satellite RNA of Bamboo Mosaic Virus

Satellite RNA of BaMV is another example of a relatively large satellite RNA that can function as an mRNA for a nonstructural protein. Like other potexviruses, BaMV has a 6.4-kb single-stranded RNA genome, and infected cells contain two unencapsidated subgenomic BaMV-specific RNAs of 2.1kb and 1.0kb. However, RNA extracted from virus particles purified from plants infected with certain isolates of BaMV (Lin and Hsu 1994) contained extra RNA molecules. These were linear single-stranded RNA molecules of 836 nt that, like the genome RNA of BaMV, were linked to a 3'-terminal poly(A) tail. Apart from polyadenylation, this RNA had no marked sequence similarities with BaMV RNA; the only similar features were the six 5'-terminal nucleotides GAAAAC, some other common stretches in the 5'-noncoding region, and the sequence ACCUAA near the 3' terminus, features also present in RNA genomes of other potexviruses (Lin and Hsu 1994). The 836-nt BaMV-associated RNA is completely dependent on BaMV for its replication but is not required for the multiplication of BaMV. Thus, by definition, this RNA is a satellite RNA. At present, this is the only satellite RNA of viruses in the genus *Potexvirus* , or indeed of any viruses with filamentous particles. BaMV satellite RNA, like other satellite RNAs, is encapsidated by the coat protein of its helper virus, but unlike all other known satellite RNAs it is encapsidated to form filamentous particles. The length of the particles (60 nm) corresponds to the size of the satellite RNA (Lin and Hsu 1994). If, as seems likely, BaMV RNA contains an origin of assembly from which coat protein encapsidates the genomic RNA, then this origin should be duplicated in the satellite RNA. For other potexvirus RNAs, the origin of assembly is at the 5' end (AbouHaidar and Erickson 1985). Possibly, the sequence shared between BaMV satellite RNA and BaMV RNA is part or all of this origin of assembly (Lin and Hsu 1994).

BaMV satellite RNA contains an ORF for a protein of 183 amino acids that is flanked by noncoding regions of 159nt (5') and 129nt (3'). This ORF can be translated in vitro (Lin and Hsu 1994), and the translation product has been found in infected plants (Lin et al. 1996). The translation product migrates in electrophoresis like an M_r 25 kDa protein rather than at the rate predicted from the size of the ORF (Lin and Hsu 1994; Lin et al. 1996). This discrepancy may be due to its highly basic character. The hydropathy profile of this protein shows it to have a highly hydrophilic domain at the N-terminus, followed by hydrophobic stretches of amino acids.

In contrast to the protein translation products of nepovirus satellite RNAs (see above), the protein encoded by BaMV satellite RNA is not essential for its replication. Mutants containing an ORF coding for this protein, but altered to eliminate its translation, were able to replicate in barley protoplasts (LIN et al. 1996) and in inoculated *C. quinoa*. Moreover, precise replacement of the ORF with sequences encoding bacterial chloramphenicol acetyltransferase resulted in a high level of expression of the foreign gene, which suggested that the modified satellite RNA is still able to replicate. Thus, the function of the 20 kDa protein remains to determined. No significant homology was detected with proteins encoded by other large satellite RNAs, although they have some properties in common, such as a strong net positive charge in the N-terminal region. The presence of arginine-rich sequences at the N termini in these proteins might suggest a common RNA-binding motif that may assist in replication, movement, or other functions of satellite RNA. Database searches with the amino acid sequence of the 20-kDa protein revealed little, except for a 46% identity score with the coat protein of satellite virus associated with panicum mosaic virus (LIU and LIN 1995).

BaMV satellite RNA did not replicate when co-inoculated with potato virus X, tobacco mosaic virus, or cucumber mosaic virus (LIN and HSU 1994). The presence of BaMV satellite RNA caused a reduction of 65–85% in the accumulation of the genomic BaMV RNA in infected barley protoplasts. However, the effect of this satellite on BaMV symptom expression remains obscure.

3 Satellite-like RNAs

3.1 Groundnut Rosette Virus Satellite-like RNA

Rosette disease of groundnuts occurs widely in Africa south of the Sahara and causes severe crop damage. The disease is caused by a complex of agents that consists of two viruses and a satellite-like RNA. GRV is a member of genus *Umbravirus*, which contains viruses that do not form conventional virus particles and depend on a helper virus, usually a luteovirus, for transmission by aphids (MURANT et al. 1995; TALIANSKY et al. 1996). For transmission of GRV, the helper virus is groundnut rosette assistor virus (GRAV; family *Luteoviridae*), but neither it nor GRV can individually induce the symptoms of groundnut rosette disease in groundnuts (HULL and ADAMS 1968; REDDY et al. 1985). The third component of the disease complex is a single-stranded RNA, 895–903 nt long. This RNA plays a key role in the induction of the symptoms of groundnut rosette disease (MURANT et al. 1988; MURANT and KUMAR 1990). This RNA relies on GRV for its replication, but more unusually, it is needed (together with GRAV) for GRV to be aphid-transmissible (MURANT 1990). Thus this ~900 nt RNA is essential for the survival of GRV in nature. It is therefore of uncertain status according to the definition of satellites given in the 6th ICTV Report (Mayo et al. 1995). The 7th ICTV Report will be less equivocal, and the definition given for satellites excludes agents that contribute an essential function to other RNA components.

Nucleotide sequences have been determined for ten clones representing five variants of the ~900 nt RNA associated with isolates that induce different types of symptoms in groundnuts or in the experimental host *Nicotiana benthamiana* (Blok et al. 1994). The sequence of GRV satellite-like RNA has no significant homology with GRV genome RNA except for very short (no longer than 10nt) 5′- and 3′-terminal regions (Taliansky et al. 1996). Different GRV satellite-like RNA variants contain up to five potential ORFs in positive or negative sense (Blok et al. 1994). However, none of the initiation codons are in good contexts for translation and no translation products have been detected in vitro (A. Ziegler, unpublished). Moreover, results of experiments in which site-directed mutagenesis was used to replace the initiation codon of each ORF with another triplet, as well as of experiments with deletion mutants, have shown that none of the ORFs is essential for replication or spread of the satellite-like RNAs. Also, production of yellow blotch symptoms by the YB variant of satellite-like RNA in *N. benthamiana* did not require any of the ORFs to be translatable but involved two elements in the satellite-like RNA, designated elements A and B, that could act *in trans* (Fig. 2; Taliansky and Robinson 1997a). It should be noted that the symptoms induced by satellite-like RNA from different GRV isolates are independent of the helper GRV isolate, and indeed indistinguishable symptoms were induced when a different 'helper' virus, PEMV, was substituted for GRV (Demler et al. 1996a). Thus, it is possible that GRV satellite-like RNA may induce symptoms itself without any specific contribution from GRV helper.

Normally, GRV satellite-like RNA does not affect the accumulation of GRV genomic and subgenomic RNAs in infected plants, but a few so-called mild GRV satellites have been identified (Murant and Kumar 1990). One such variant, NM3c, which induces very mild symptoms, has been shown to diminish drastically the replication of the helper GRV in infected or transgenic *N. benthamiana* plants (Taliansky and Robinson 1997b; Taliansky et al. 1998). This effect on replication did not involve any of the satellite ORFs but was controlled by a region near its 5′

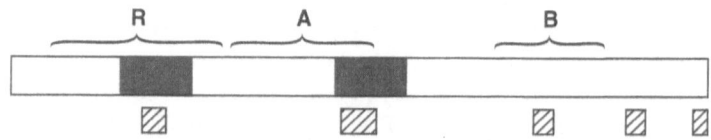

Fig. 2. Functional elements of the 900-nt GRV satellite-like RNA. The *bar* represents the GRV satellite-like RNA. *R* indicates the location of the *cis*-acting element required for replication, and *A* and *B* indicate the locations of the *trans*-acting RNA elements involved in symptom induction. The R-element of mild GRV satellite-like RNA NM3c is also involved in down-regulation of GRV replication. The locations of the duplicated sequences are indicated by *shading*. *Cross-hatched boxes* indicate where sequences of > 20 contiguous nucleotides are 80% or more identical with sequences in PEMV satellite RNA

end (R-element), which is required in *cis* for replication of GRV satellite-like RNA (Fig. 2; TALIANSKY and ROBINSON 1997a,b).

As mentioned above, the presence of GRV satellite-like RNA is required for GRAV-dependent transmission of GRV (MURANT 1990). Although it cannot be completely ruled out that a mechanism for this requirement is mediated by one of the potential proteins encoded by the satellite-like RNA, this suggestion seems to be unlikely (see above). It is more likely that satellite-like RNA has a structural function, such as the stabilization of virus particles containing GRV RNA in a coat of GRAV capsid protein.

The nucleotide sequence of satellite-like RNA of GRV resembles that of satellite RNA associated with PEMV (DEMLER et al. 1996b). Both satellites contain internal repetitions, although their significance is unknown. PEMV satellite RNA contains two 27-nt exact repeats and GRV satellite-like RNA contains two longer but imperfect duplications of 93 and 94nt (Fig. 2; DEMLER et al. 1996b). However, in contrast to the ~900-nt RNA associated with GRV, the 717-nt satellite RNA of PEMV had no detectable effect on symptoms induced by PEMV in its natural host, *Pisum sativum* , although it did attenuate the symptoms induced in *N. benthamiana* (DEMLER et al. 1996b). Moreover, PEMV satellite had no effect on the aphid transmissibility of PEMV. Thus PEMV satellite is fully dispensable, like a classical satellite (MAYO et al. 1995). However, GRV and PEMV are each able to support the replication and systemic spread of homologous and heterologous satellites (DEMLER et al. 1996a), which implies a close relationship between the satellites. Thus, the large satellite-like RNA of GRV may represent some intermediate class of satellites (still large but without messenger activity). Moreover, it could be thought to be evolving towards becoming an essential genomic RNA.

3.2 Ancilliary RNA of Beet Necrotic Yellow Vein Virus

Infection of sugar beet crops with BNYVV results in a serious disease of the roots, termed rhizomania, which describes the massive root proliferation symptom induced by the infection. Virus is transmitted in soil by zoospores of the fungus, *Polymyxa betae* . The rod-shaped particles of BNYVV are of several lengths and contain four or five RNA components. The smallest RNA (RNA-5) is found in a number of Japanese isolates (TAMADA et al. 1989; KIGUCHI et al. 1996) and in occasional isolates from Europe (KOENIG et al. 1997). All RNA components share the same nucleotide sequence for 70 nucleotides upstream of the 3'-terminal poly(A) sequence. In contrast, sequence homology at the 5' ends of the RNAs is limited to a few nucleotides.

Isolates of BNYVV that had been kept in laboratory culture in *C. quinoa* or *Tetragonia expansa* and transmitted from plant to plant by mechanical inoculation, were found sometimes to contain RNA-3 and/or RNA-4 that were shorter than those in virus cultures freshly isolated from the field. Some isolates were found to lack one of the smaller RNA components altogether. RNA-1 and RNA-2 were always present, and it has been concluded that BNYVV has a bipartite genome with RNA-1 and RNA-2 responsible for the "house-keeping functions" of the virus (Jupin et al. 1991). Nucleotide sequencing and mutagenesis work has shown that these two RNAs contain genes with putative replication, encapsidation, virus movement, and fungus transmission functions.

RNA-3, RNA-4, and RNA-5 each contain one major ORF (Fig. 3) and seem to resemble satellite RNAs. However, the composition of field isolates suggests that these RNAs contribute to virus survival, and mutagenesis work has shown that each contributes one or more useful properties for virus success. Thus these RNAs may better be classified as satellite-like RNAs.

The presence of RNA-3 can result in marked changes in the symptoms that RNA-1 and RNA-2 alone can induce in *C. quinoa* or *T. expansa* . Without RNA-3, inocula induced faint lesions and weak chlorotic spots or rings. When RNA-3 was present, the lesions were bright yellow (Tamada et al. 1989). RNA-3 encodes the polypeptide P25 (Fig. 3); mutagenesis that prevented the expression of P25 also eliminated the yellowing effect on the local lesions (Jupin et al. 1992). The symptoms induced in *Beta macrocarpa* were also more severe when RNA-3 was present, and this effect was most marked in the roots. In infected sugar beet, when RNA-3 was absent the yield losses from infection were slight, but when RNA-3 was present, the losses reached 95% (Koenig et al. 1991). The yield loss was correlated with an increased virus concentration in tap roots, which suggests that RNA-3

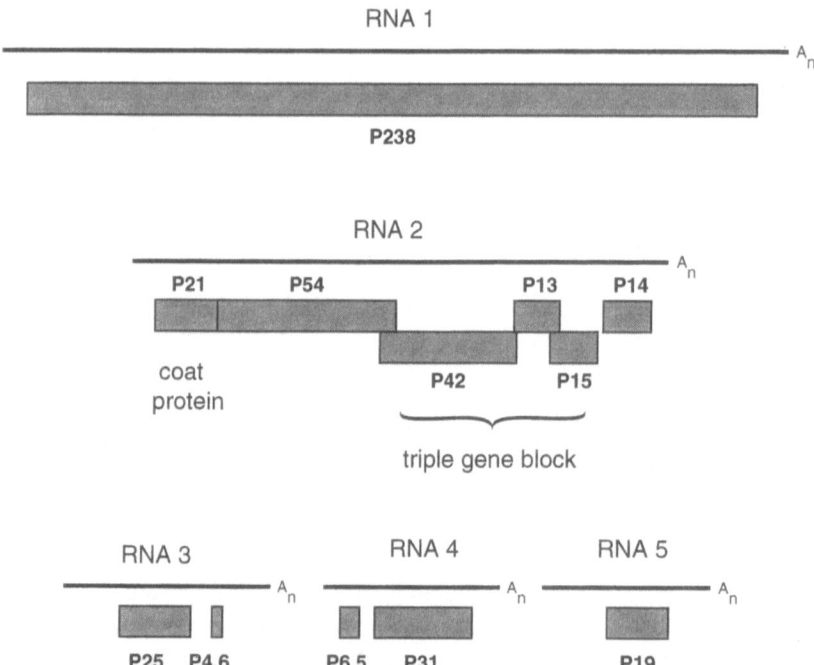

Fig. 3. The genome and satellite RNAs of beet necrotic yellow vein virus. *Solid lines* represent the RNA molecules; *shaded boxes* represent the ORFs

facilitates multiplication and spread in the roots. Also, RNA-3 was essential for movement of virus into the vascular system of *B. macrocarpa* (TAMADA et al. 1989). Recent results have shown that mutant RNA-3, in which translation of the P25 gene and the other putative ORFs was prevented nonetheless functioned to allow movement of virus in the vascular system; the functional domain is at least in part a property of the nucleotide sequence rather than of the protein (LAUBER et al. 1998). Thus RNA-3 contributes in large measure to the pathogenicity of the virus, although the details of the mechanism are unknown (LAUBER et al. 1998).

The presence of RNA-4 can slightly increase the degree of chlorosis in infected plants, and more so if RNA-5 is also present (TAMADA et al. 1989). The major effect of RNA-4 is that it greatly increases the efficiency of transmission of the virus by the fungus vector. In effect, isolates are barely pathogenic if RNA-4 is absent.

RNA-5 seems to play a role in both symptom production and fungus transmission. Isolate D5 (containing RNA-1 + RNA-2 + RNA-5) (KIGUCHI et al. 1996) was found to induce chlorotic lesions (mild symptoms), whereas isolate D6 (containing RNA-1, RNA-2, and a deleted form of RNA-5 that abolished the major ORF) induced only faint chlorotic lesions (very mild symptoms). Also, isolate D5 was able, on occasion, to infect *B. macrocarpa* systemically (as it does when RNA-3 is present in isolates) but isolate D6 was not (TAMADA et al. 1989). These data suggest that the P26 translation product of RNA-5 has a role in the induction of symptoms.

In the absence of RNA-3, the presence of RNA-5 induces scab-like symptoms instead of typical rhizomania (stunting and abnormal proliferation of the rootlets). Interestingly, RNA-3 and RNA-5 have synergetic effects in susceptible cultivars, and isolates containing RNA-5 but lacking RNA-3 cause more severe damage in resistant cultivars than in susceptible cultivars. Therefore, RNA-3 and RNA-5 play different roles in symptom development in sugar-beet roots.

KOENIG et al. (1997) observed that rhizomania is particularly severe in the area infected by P-type BNYVV, which contains RNA-5. P25 (RNA-3) and P26 (RNA-5) share a conserved amino acid sequence (FRGPGN). This sequence may be involved in pathogenecity, because a deletion mutant which lacks a coding region including this motif has lost its specific pathogenicity for *C. quinoa* and *B. macrocarpa* . Also, isolates that contained both RNA-4 and RNA-5 were transmitted more efficiently by *P. betae* than an isolate containing RNA-4 alone. Thus, RNA-5 can also assist the efficiency of transmission, and may therefore add to, or partially replace, the effects of RNA-3 and RNA-4.

4 Discussion

The different satellite-like systems described in this chapter illustrate the extensive overlap between what might at first have been considered to be easily discriminated, namely parts of a virus genome and satellites that 'parasitize' virus genomes.

Figure 4 illustrates this range of overlapping systems. The extremes are well separated, but between them are a variety of intermediate associations. These are the satellite RNAs that utilize helper virus coat protein, either the small non-mRNAs or the larger mRNAs. BaMV satellite is a further intermediate, in that the encoded protein seems to have little to do with satellite multiplication. Satellite-like RNAs are a further stage towards dependency, in that they contribute valuable functions to the helper virus. Where the helping component is itself an independent virus, the relationship is simply that of a defective virus depending on another virus. An example is PVC that lacks functional helper protein but can be transmitted from tissues containing the helper protein of PVY because the heterologous protein is functional (KASSANIS and GOVIER 1971). Similarly, RTBV and AYV can provide helper protein needed for the vector transmission of RTSV and PYFV, respectively (HARRISON and MURANT 1984). This trend is further developed in the example of

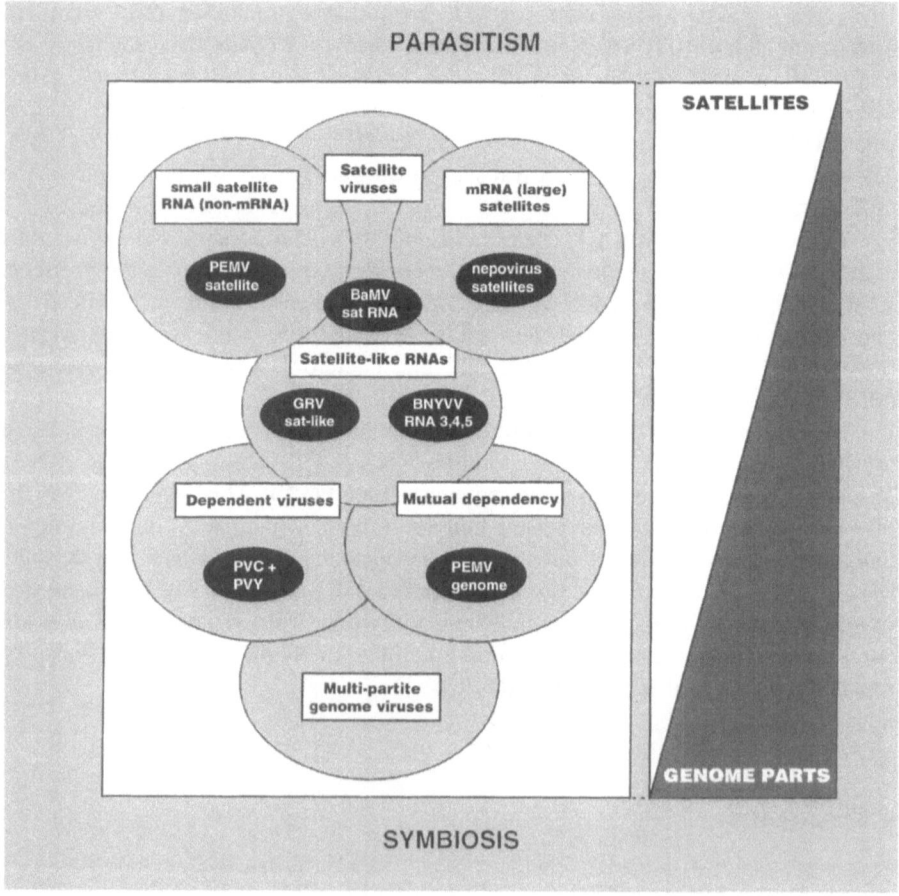

Fig. 4. Stages of the imagined evolution of RNA segments to, or away from, dependency, i.e., between satellites (parasitism) and genome parts (symbiosis). *Large gray zones* indicate the different stages, *black lozenges* indicate examples referred to in the text

PEMV, in which each RNA component encodes a functional replicase, but only the combination of the two RNAs is capable of survival and spread in nature (DEMLER et al. 1996b). The final stage of the supposed gradient is the classical multipartite genome, where only one component encodes replicase. Although this component can support multiplication of the RNA, the virus can multiply only when all RNAs are present.

The conclusion, based on any consideration of these overlapping systems, and especially the 'benign' satellites, must be that, as is often true in biology, attempts to categorize agents in discrete types (genome part, satellite, symbiotic RNA, etc.) are always going to fail. RNA molecules will evolve into, and presumably out of, dependency, and viruses will sometimes evolve to exploit a phenotype contributed by an associated RNA in preference to that encoded in their genomes. During such an evolutionary process, it is going to be impossible to define when a molecule is a commensal, when it is a parasite, when it is a symbiont, and when it becomes a part of the genome of the virus. And, as we attempt to demonstrate in Fig. 4, evolution is equally imaginable in either direction.

Mutual dependency between RNA components has been described as reflecting the multipartite nature of the genome of the virus (i.e., a virus is the sum of its parts). But for one such virus (pea enation mosaic virus), recent taxonomic proposals are to divide and classify the parts of the 'genome' as separate viruses in different genera, largely because each RNA molecule encodes its own polymerase (D'ARCY and MAYO 1997). However, in nature, only the complex of the two agents can survive. In other examples, one virus can supply a gene *in trans* to complement an otherwise defective virus. An extreme example is that the replicase-coding parts of the bipartite genomes of comoviruses, dianthoviruses, nepoviruses and tobraviruses can multiply independently of the smaller genome part (MAYO 1992, and references therein). For tobraviruses, the defective part can even cause a disease (HARRISON and ROBINSON 1986). But the missing genome part supplies a gene product essential for the survival of the virus in natural infections, either the coat protein (tobraviruses, comoviruses, nepoviruses) or movement protein (dianthoviruses). Thus, none of these RNA, like the parts of PEMV, or any of the satellite RNA, could survive alone in nature. The existence of 'benign' satellites and similar RNAs prompts the philosophical question: "What is a virus?" No answer is offered here.

References

AbouHaidar MG, Erickson JW (1985) Structure and *in vitro* assembly of papaya mosaic virus. In: Davies JW (ed) Molecular plant virology, vol 1. CRC Press, Boca Raton, pp 85 121

Baulcombe DC, Saunders GR, Bevan MW, Mayo MA, Harrison BD (1986) Expression of biologically active viral satellite RNA from the nuclear genome of transformed plants. Nature 321:446 449

Blok VC, Ziegler A, Robinson DJ, Murant AF (1994) Sequences of 10 variants of the satellite-like RNA3 of groundnut rosette virus. Virology 202:25 32

Bouzoubaa S, Guilley H, Jonard G, Richards K, Putz C (1985) Nucleotide sequence analysis of RNA-3 and RNA-4 of beet necrotic yellow vein virus, isolates F2 and G1. J Gen Virol 66:1553 1564

D'Arcy CJ, Mayo MA (1997) Proposals for changes in luteovirus taxonomy and nomenclature. Arch Virol 142:1285-1287

Demler SA, Rucker DG, Nooruddin L, de Zoeten GA (1994) Replication of the satellite RNA of pea enation mosaic virus is controlled by RNA 2-encoded functions. J Gen Virol 75:1399 1406

Demler SA, Rucker DG, de Zoeten GA, Ziegler A, Robinson DJ, Murant AF (1996a) The satellite RNAs associated with the groundnut rosette disease complex and pea enation mosaic virus: sequence similarities and ability of each other's helper virus to support their replication. J Gen Virol 77:2847 2855

Demler SA, De Zoeten GA, Adam G, Harris KF (1996b) Pea enation mosaic enamovirus: properties and aphid transmission. In: Harrison BD, Murant AF (eds) The plant viruses, vol 5: polyhedral virions and bipartite genomes, chap 12. Plenum, New York, pp 303 334

Fritsch C, Mayo MA, Murant AF (1978) Translation of the satellite RNA of tomato black ring virus *in vitro* and in tobacco protoplasts. J Gen Virol 40:587 593

Fritsch C, Mayo MA, Murant AF (1980) Translation products of genome and satellite RNAs of tomato black ring virus. J Gen Virol 46:381-389

Fritsch C, Koenig I, Murant AF, Raschke JH, Mayo MA (1984) Comparisons among satellite RNA species from five isolates of tomato black ring virus and one isolate of myrobalan latent ringspot virus. J Gen Virol 65:289-294

Fritsch C, Mayo MA, Hemmer O (1993) Properties of the satellite RNA of nepoviruses. Biochimie 75:561-567

Fuchs M, Pinck M, Serghini MA, Ravelonandro M, Walter B, Pinck L (1989) The nucleotide sequence of satellite RNA in grapevine fanleaf virus, strain F13. J Gen Virol 70:955-962

Gallitelli D, Savino V, De Sequeira OA (1983) Properties of a distinctive strain of grapevine Bulgarian latent virus. Phytopathol Medit 22:27 32

Hans F, Pinck M, Pinck L (1993) Location of the replication determinants of the satellite RNA associated with grapevine fanleaf nepovirus (strain F13). Biochimie 75:597 603

Harrison BD, Murant AF (1984) Involvement of virus-coded proteins in transmission of plant viruses by vectors. In: Mayo, MA, Harrap KA (eds) Vectors in virus biology. Academic Press, New York, pp 1 36

Harrison, BD, Robinson DJ (1986) Tobraviruses. In: Van Regenmortel MHV, Fraenkel-Conrat H (eds) The plant viruses, vol 2. Plenum Press, New York, pp 339 369

Harrison BD, Mayo MA, Baulcombe DC (1987) Virus resistance in transgenic plants that express cucumber mosaic virus satellite RNA. Nature 328:799 802

Hemmer O, Meyer M, Greif C, Fritsch C (1987) Comparison of the nucleotide sequences of five tomato black ring virus satellite RNAs. J Gen Virol 68:1823-1833

Hemmer O, Oncino C, Fritsch C (1993) Efficient replication of the *in vitro* transcripts from cloned cDNA of tomato black ring virus satellite RNA requires the 48K satellite RNA-encoded protein. Virology 194:800 806

Hull R, Adams AN (1968) Groundnut rosette and its assistor virus. Ann Appl Biol Commun 151:388 395

Jupin I, Tamada T, Richards K (1991) Pathogenesis of beet necrotic yellow vein virus. Semin Virol 2:121 129

Jupin I, Guilley H, Richards KE, Jonard G (1992) Two proteins encoded by beet necrotic yellow vein virus RNA 3 influence symptom phenotype on leaves. EMBO J 11:479 488

Kassanis B (1962) Properties and behaviour of a virus depending for its multiplication on another. J Gen Microbiol 27:77-488

Kassanis B, Govier DA (1971) New evidence on the mechanism of aphid transmission of potato aucuba mosaic virus and potato virus C. J Gen Virol 10:99 101

Kiguchi T, Saito M, Tamada T (1996) Nucleotide sequence analysis of RNA-5 of five isolates of beet necrotic yellow vein virus and the identity of a deletion mutant. J Gen Virol 77:575 580

Koenig R, Jaraush W, Li Y, Commandeur U, Burgermeister W, Gehrke M, Lüddecke P (1991) Effect of recombinant beet necrotic yellow vein virus with different RNA compositions on mechanically inoculated sugarbeets. J Gen Virol 72:2243-2246

Koenig R, Haeberlé AM, Commandeur U (1997) Detection and characterization of a distinct type of beet necrotic yellow vein virus RNA 5 in a sugarbeet growing area in Europe. Arch Virol 142:1499 1504

Kreiah S, Cooper JI, Strunk G (1993) The nucleotide sequence of a satellite RNA associated with strawberry latent ringspot virus. J Gen Virol 74:1163-1165

Lauber E, Guilley H, Tamada T, Richards KE, Jonard G (1998) Vascular movement of beet necrotic yellow vein virus in *Beta macrocarpa* is probably dependent on an RNA 3 sequence domain rather than a gene product. J Gen Virol 79:385 393

Leader DP, Katan M (1988) Viral aspects of protein phosphorylation. J Gen Virol 69:1441 1464

Lin N-S, Hsu Y-H (1994) A satellite RNA associated with bamboo mosaic potexvirus. Virology 202:707 714

Lin N-S, Lee Y-S, Lin B-Y, Lee C-W, Hsu Y-H (1996) The open reading frame of bamboo mosaic potexvirus satellite RNA is not essential for its replication and can be replaced with a bacterial gene. Proc Natl Acad Sci USA 93:3138 3142

Liu J-S, Lin N-S (1995) Satellite RNA associated with bamboo mosaic potexvirus shares similarity with satellites associated with sobemoviruses. Arch Virol 140:1511–1514

Liu YY, Hellen CUT, Cooper JI, Bertioli DJ, Coates D, Bauer G (1990) The nucleotide sequence of a satellite RNA associated with arabis mosaic nepovirus. J Gen Virol 71:1259–1263

Liu YY, Cooper JI (1993) The multiplication in plants of arabis mosaic virus satellite RNA requires the encoded protein. J Gen Virol 74:1472 1474

Liu YY, Cooper JI, Coates D, Bauer G (1991a) Biologically active transcripts of a large satellite RNA from arabis mosaic nepovirus and the importance of 5′ end sequences for its replication. J Gen Virol 72:2867–2874

Liu YY, Cooper JI, Edwards ML, Hellen CUT (1991b) A satellite RNA of arabis mosaic nepovirus and its pathological impact. Ann Appl Biol 118:577 587

Mayo MA (1992) How important is genome division as a taxonomic criterion in plant virus classification? Arch Virol [Suppl] 5:183 187

Mayo MA, Berns K, Fritsch C, Jackson AO, Kaper J, Leibowitz M, Taylor JM (1995) Satellites. In: Murphy FA, Fauquet CM, Bishop DHL, Ghabrial SA, Jarvis AW, Martelli GP, Mayo MA, Summers MD (eds) Virus taxonomy the classification and nomenclature of viruses: Sixth Report of the International Committee on Taxonomy of Viruses. Springer, Vienna New York, pp 487 492

Moser O, Fuchs M, Pinck L, Stussi-Garaud C (1992) Immunodetection of grapevine fanleaf virus satellite RNA-encoded protein in infected *Chenopodium quinoa*. J Gen Virol 73:3033 3038

Murant AF (1990) Dependence of groundnut rosette virus on its satellite RNA as well as on groundnut rosette assistor luteovirus for transmission by *Aphis crassivora*. J Gen Virol 71:2163 2166

Murant AF, Kumar IK (1990) Different variants of the satellite RNA of groundnut rosette virus are responsible for the chlorotic and green forms of groundnut rosette disease. Ann Appl Biol 117:85 92

Murant AF, Mayo MA (1982) Satellites of plant viruses. Annu Rev Phytopathol 20:49 70

Murant AF, Mayo MA, Harrison BD, Goold RA (1973) Evidence for two functional RNA species and a "satellite" RNA in tomato black ring virus. J Gen Virol 19:275 278

Murant AF, Rajeshwari R, Robinson DJ, Rascke JH (1988) A satellite RNA of groundnut rosette virus is largely responsible for symptoms of groundnut rosette disease. J Gen Virol 69:1479 1486

Murant AF, Robinson DJ, Gibbs MJ (1995) Genus *Umbravirus*. In: Murphy FA, Fauquet CM, Bishop DHL, Gabrial SA, Jarvis AW, Martelli GP, Mayo MA, Summers MD (eds) Virus taxonomy classification and nomenclature of viruses: Sixth Report of the International Committee on the Taxonomy of Viruses. Springer, Vienna New York, pp 388–391

Oncino O, Hemmer O, Fritsch C (1995) Specificity in the association of tomato black ring virus satellite RNA with helper virus. Virology 213:87 96

Pinck L, Fuchs M, Pinck M, Ravelonandro M, Walter B (1988) A satellite RNA in grapevine fanleaf virus strain F13. J Gen Virol 69:233 239

Reddy DVR, Murant AF, Duncan GH, Ansa OA, Demski JW, Kuhn CW (1985) Viruses associated with chlorotic rosette and green rosette diseases of groundnut groundnuts. Ann Appl Biol 45:318 326

Rubino L, Tousignant ME, Steger G, Kaper JM (1990) Nucleotide sequence and structural analysis of two satellite RNAs associated with chicory yellow mottle virus. J Gen Virol 71:1897 1903

Schneider IR (1969) Satellite-like particle of tobacco ringspot virus that resembles tobacco ringspot virus. Science 166:1627–1629

Taliansky ME, Robinson DJ (1997a) *Trans*-acting untranslated elements of groundnut rosette virus satellite RNA are involved in symptom production. J Gen Virol 78:1277–1285

Taliansky ME, Robinson DJ (1997b) Down-regulation of groundnut rosette virus replication by a variant satellite RNA. Virology 230:228 235

Taliansky ME, Robinson DJ, Murant AF (1996) Complete nucleotide sequence and organization of the RNA genome of groundnut rosette umbravirus. J Gen Virol 77:2335 2345

Taliansky ME, Ryabov EV, Robinson DJ (1998) Two distinct mechanisms of transgenic resistance mediated by groundnut rosette virus satellite RNA sequences. Mol Plant Microbe Interact 11:367 374

Tamada T, Shirako Y, Abe H, Saito M, Kiguchi T, Harada T (1989) Production and pathogenicity of isolates of beet necrotic yellow vein virus with different numbers of RNA components. J Gen Virol 70:3399 2409

Tien Po, Chang XH (1983) Control of two seed-borne viruses in China by the use of protective inoculation. Seed Sci Tech 11:969 972

Encapsidated Circular Viroid-like Satellite RNAs (Virusoids) of Plants

R.H. Symons[1] and J.W. Randles[2]

1 Introduction

The first encapsidated circular, viroid-like satellite RNA of a plant virus was described by RANDLES et al. (1981). In less than a year, three others were identified (FRANCKI 1987) but, strangely, it was not until 17 years later that a fifth one was reported (COLLINS et al. 1998). The five viruses which harbour these circular satellite RNAs are listed in Table 1; all are members of the sobemovirus, or Southern bean mosaic virus, group (FRANCKI 1987) which contains at least the nine members listed in Tables 1 and 2. Another unusual feature is that the viruses supporting these satellite RNAs have been found only in Australia and New Zealand (Australasia), with the exception of one isolate of lucerne transient streak virus (LTSV)

[1] Department of Plant Science, Waite Institute, University of Adelaide, Glen Osmond, SA 5064, Australia
[2] Department of Crop Protection, Waite Institute, University of Adelaide, Glen Osmond, SA 5064, Australia

Table 1. Sobemoviruses containing virusoids

Virus	Length (nt)	
	Viral RNA	Virusoid
Lucerne transient streak virus (LTSV)	~4275[a]	322, 324[b]
Rice yellow mottle virus (RYMV)	4450[c]	220[d]
Solanum nodiflorum mottle virus (SNMV)		377[e]
Subterranean clover mottle virus (SCMoV)	–	332, 388[f]
Velvet tobacco mottle virus (VTMoV)	–	365, 366[e]

[a] Jeffries, Rathjen, and Symons, unpublished.
[b] KEESE et al. (1983); ABOUHAIDAR and PALIWAL (1988).
[c] YASSI et al. (1994).
[d] COLLINS et al. (1998).
[e] HASELOFF and SYMONS (1982).
[f] DAVIES et al. (1990).

Table 2. Sobemoviruses not known to contain specific virusoids

Virus	Length of Viral RNA (nt)
Cocksfoot mottle virus (CfMV, CoMV)	4082/4083[a]
Southern bean mosaic virus (SBMV)	Bean strain (B) 4109[b]
	Cowpea strain (C) 4194[c]
Sowbane mosaic virus (SoMV)	–
Turnip rosette virus (TRoSV)	–

[a] MÄKINEN et al. (1995).
[b] OTHMAN and HULL (1995).
[c] WU et al. (1987).

reported from Canada (PALIWAL 1983) and rice yellow mottle virus (RYMV) found in Africa (YASSI et al. 1994).

These encapsidated circular RNAs vary in length from 324 to 388 nucleotides and will be called virusoids throughout this chapter. This term was first used in 1982 (HASELOFF et al. 1982) because of the viroid-like structure and size of these RNAs and has been widely used since. It has practical advantages in brevity and in serving to distinguish the virusoids from the satellite RNA of tobacco ringspot virus (sTRSV), which occurs in both circular and linear forms in vivo, but only the linear form of which is encapsidated. In addition, TRSV is a member of the nepovirus group.

2 History of Discovery

Velvet tobacco mottle virus (VTMoV) was found in May 1979 on native velvet tobacco (*Nicotiana velutina*) during an expedition by one of the authors (JWR) and

C. Davies to collect isolates of tobacco mosaic virus (TMV) from tree tobacco (*N. glauca*) in the arid zone of South Australia, north of Adelaide. The sites where it was found were Cobblers Sandhill (29°26′, 139°57′) (Fig. 1) and Coopers Creek near Innamincka (27°26′, 140°42′). Rugose mottling was seen on many plants, and subsequent transfer to experimental hosts showed that the virus was unrelated to TMV. The virus was transferred to and readily purified from *N. clevelandii* and was shown to be a 30-nm icosahedron, stable, and highly immunogenic. In an attempt to determine the size of the viral RNA by electron microscopy, long linear molecules were observed, but circular molecules about the size of viroids were also prevalent (Fig. 2).

Another unusual property was the ability of VTMoV to be naturally transmitted by a mirid bug (*Cyrtopetis nicotianae*); this is still the only example of virus transmission by a mirid. No nematode vector species were found in soil samples at the field site, and testing of a few individuals of other hemipterous insect species trapped at the site failed to show that they transmitted VTMoV. The plant species associated with velvet tobacco at Cobbler's Sandhill, *N. glauca* and *Phyllanthus lacunarius* (*Euphorbiaceae*) were not infected. The virus has been found to be prevalent in velvet tobacco on several subsequent visits to this site and has also been isolated from velvet tobacco at Cunnamulla in southwestern Queensland (G. Behncken, unpublished).

The biological properties of VTMoV suggested that it was unique, but when a comparison was made with antigen and antiserum collections held in the UK by

Fig. 1. Cobbler's Sandhill on the Strezelecki Track in the northeastern region of South Australia, where VTMoV-infected *Nicotiana velutina* were first found. The botanical composition also includes *N. glauca* (tree tobacco) and *Phyllanthus lacunarius*

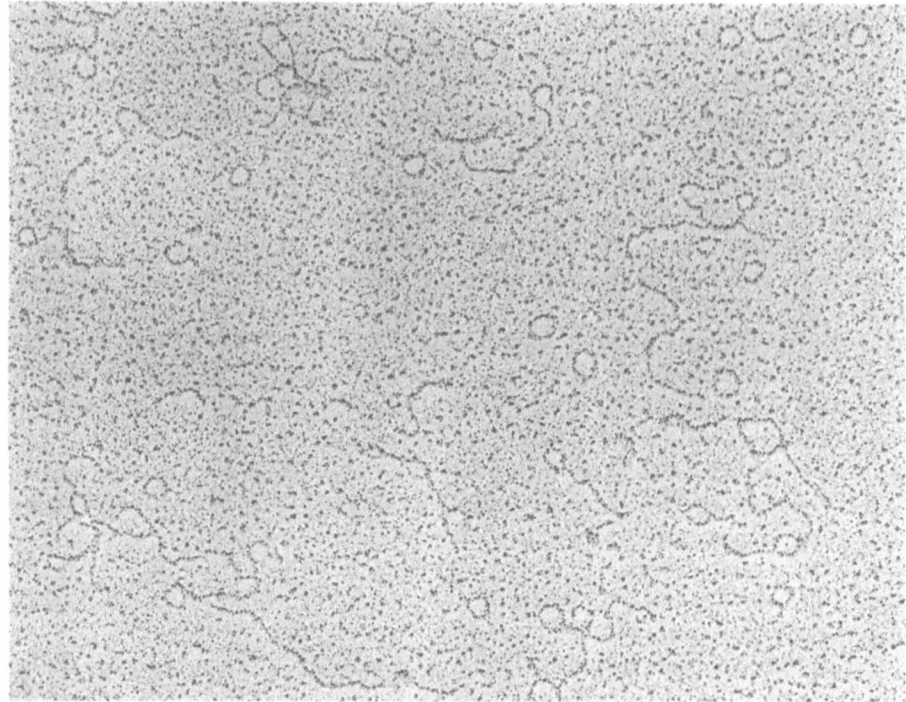

Fig. 2. Electron micrograph of total RNA isolated from VTMoV, spread under denaturing conditions, and shadowed at 5° with Pt-palladium. Note the prevalent small circular RNAs and the longer linear RNAs, which probably represent the genomic RNAs shown in Fig. 3. ×80,000

Dr. M. Hollings, Glasshouse Crops Research Institute, VTMoV was found to be serologically related to the beetle-transmitted solanum nodiflorum mottle virus (SNMV), which had been previously isolated in Queensland by R.S. GREBER (1973). It was not related to any other viruses in Hollings' extensive collection. Exchange of material with R.S. Greber led to the discovery that, although VTMoV and SNMV were biologically distinct and showed only about 20% homology by hybridization analysis of their genomic RNAs, SNMV contained a circular RNA of similar size and sequence to that of VTMoV. Each of the viruses was transmitted experimentally by its reciprocal vector (RANDLES and FRANCKI 1986; GREBER and RANDLES 1986).

It seems possible that VTMoV and SNMV have a common ancestor, whose host plant may have occupied a larger area when the climate in central Australia was less arid and there was a continuum of temperate vegetation between eastern and central Australia. With climatic change, and the desertification of parts of Australia, it seems possible that plants infected with this ancestor may have been separated geographically and that different selection pressures may then have led to the speciation of the ancestor into VTMoV and SNMV and adaptation to different vectors. The origin and function of the circular RNAs in VTMoV and SNMV are unknown. They function as satellite RNAs, and perhaps they confer some selective advantage on the virus. The demonstration that VTMoV can encapsidate potato spindle tuber viroid RNA (FRANCKI et al. 1986a) raises the possibility that viruses with circular satellite RNAs may have resulted from the chance encapsidation of a viroid-like RNA from a host plant. Nevertheless, their unusual properties and the limited number of places where such circular satellites have been found indicates that they are not essential components of the virus. Nevertheless, as discussed below, they have provided a useful model for analyzing RNA replication and studying ribozymes.

3 General Properties

The sobemoviruses are small, spherical viruses which contain one molecule of single-stranded genomic RNA of about 4000–4500 nucleotides (FRANCKI 1987). VTMoV, the first sobemovirus found to also contain a virusoid, was discovered as described above, while SNMV was originally isolated from *Solanum nodiflorum*, a common weed growing along the northeastern coast of Australia (GREBER 1981) and subsequently shown to also contain a virusoid (GOULD and HATTA 1981).

Another virusoid was found in virions of LTSV (TIEN et al. 1981), a virus originally isolated from *Medicago sativa* (lucerne) in southeastern Australia (BLACKSTOCK 1978). A New Zealand isolate (FORSTER and JONES 1979) was also shown to contain the virusoid, while PALIWAL (1983) confirmed its presence in a Canadian isolate of LTSV. The fourth virusoid discovered at this time was isolated from subterranean clover mottle virus (SCMoV) infecting two clover species (*Trifolium subterraneum* and *T. globeratum*) in southwestern Australia (FRANCKI et al. 1983). The fifth and smallest virusoid was reported associated with RYMV (YASSI et al. 1994) and shown to be a virusoid by COLLINS et al. (1998).

The presence of virusoids in total viral RNA is readily shown by polyacrylamide gel electrophoresis. For example, analysis of VTMoV RNA showed the presence of three major RNA species on gel electrophoresis under denaturing conditions (Fig. 3; GOULD 1981); RNA 1 corresponds to the linear viral genomic RNA, RNA 2 to the circular virusoid, and RNA 3 to linear virusoid, as clearly indicated by electron microscopy (Fig. 3; GOULD and HATTA 1981; FRANCKI 1987). RNAs 1a and 1b presumably represent subgenomic RNAs.

4 Virusoids Are Encapsidated Circular Satellite RNAs

A satellite RNA requires a helper virus for its replication and encapsidation (MURANT and MAYO 1982; FRANCKI 1985). Although early results indicated that VTMoV and SNMV and their respective virusoids were dependent on each other

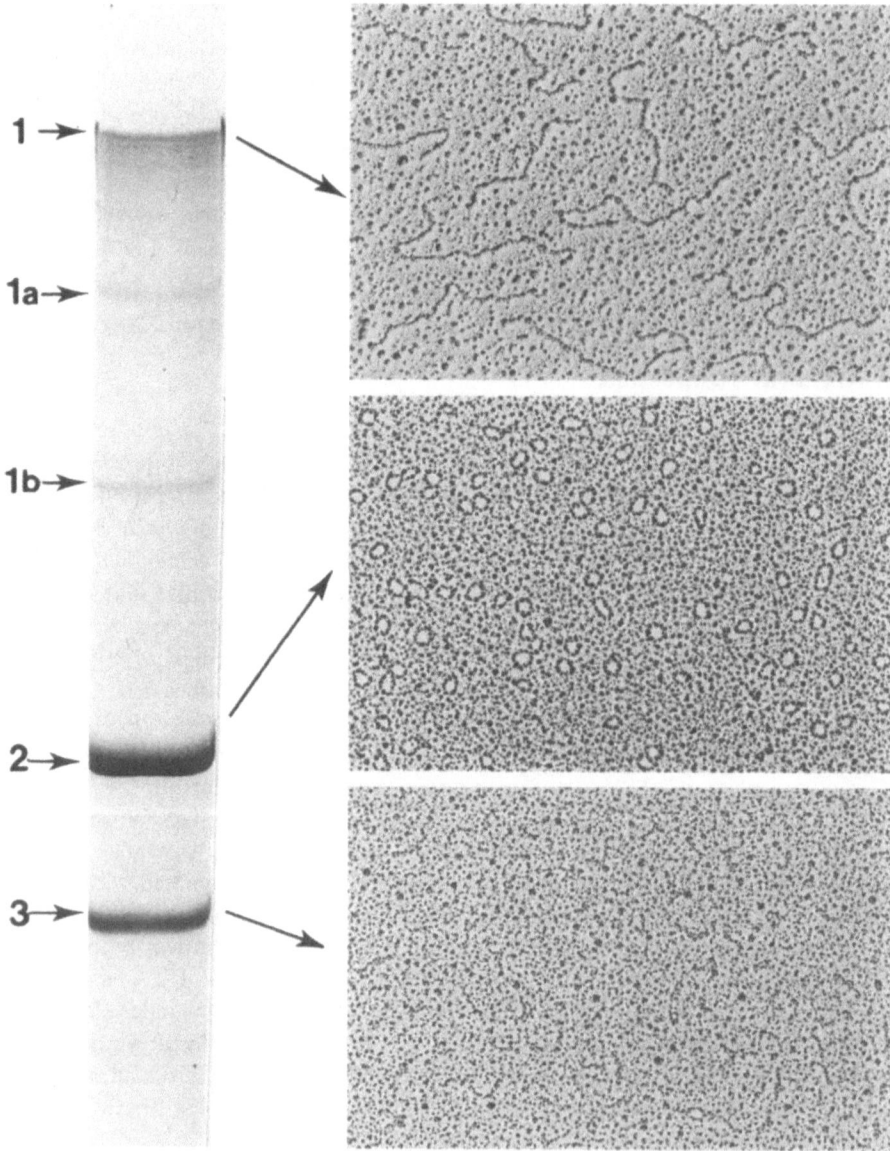

Fig. 3. Properties of the encapsidated RNAs of VTMoV. RNA isolated from purified virions was fractionated by gel electrophoresis under denaturing conditions, as shown on the *left*. Elution of RNAs 1, 2, and 3 and analysis by electron microscopy showed RNA 1 as long, linear, single-stranded RNA, RNA2 as covalently closed circles (vVTMoV), and RNA 2 as linear vVTMoV. (Reproduced from FRANCKI 1987, with permission)

for replication, i.e., the virusoids were not satellite RNAs but existed in a bipartite relationship (GOULD et al. 1981), subsequent work provided definitive data on the satellite nature of all virusoids. In the case of LTSV, JONES et al. (1983) showed that LTSV RNA could replicate autonomously but that the virusoid could not. An isolate of LTSV was maintained free of virusoid for many passages; it produced chlorotic lesions on inoculated leaves of *Chenopodium amaranticolor* whereas, on subsequent addition of virusoid, necrotic lesions were obtained (JONES et al. 1983). Further evidence of the satellite nature of vLTSV was shown by its ability to be helped by either sowbane mosaic virus (FRANCKI et al. 1983) or southern bean mosaic virus (PALIWAL 1984), both of which belong to the sobemovirus genus but are serologically unrelated to LTSV (FRANCKI 1987).

In the case of VTMoV, FRANCKI et al. (1986b) obtained virus free of virusoid and this isolate was maintained as such after passaging six times by mechanical inoculation. The systemic mosaic produced on *Nicotiana clevelandii* by this isolate also gave necrotic lesions on inoculated leaves after addition of the virusoid. With SNMV, JONES and MAYO (1983, 1984) also obtained a virusoid-free isolate. In addition, they showed that vSNMV could be helped by LTSV as well as SNMV but that vLTSV could not be helped by SNMV.

These results and those described above for vLTSV show that there is not a mutually exclusive relationship between one virusoid and one helper virus, but that there is some flexibility possible in the interrelationships. Other examples include the support by LTSV of the replication of vSCMoV under glasshouse conditions in *Chenopodium quinoa* (KEESE et al. 1983) and under field conditions in lucerne (*Medicago sativa*; DALL et al. 1990), and the support of the replication of vLTSV by cocksfoot mottle sobemovirus and turnip rosette sobemovirus (SEHGAL et al. 1993).

5 Sequence and Physical Properties

The sequences of all five virusoids have been determined and secondary structures predicted. These are presented in Fig. 4 for vVTMoV, vSNMV, v(388)SCMoV, v(332)SCMoV and vLTSV, while the sequence and predicted secondary structure of the recently described vRYMV is given in COLLINS et al. (1998). Sequencing of vVTMoV revealed two sequence variants, one of 366 nucleotides and the other with 367; the extra residue is a U in position 108 (HASELOFF and SYMONS 1982). vSNMV of 378 nucleotides has 93% sequence homology with vVTMoV (HASEL-OFF and SYMONS 1982); these two virusoids should therefore really be considered sequence variants of one virusoid rather than two separate virusoids, if the rule of thumb is accepted that greater than 90% sequence homology indicates common identity. Sequence analysis of more isolates of vVTMoV and vSNMV is required to confirm this.

Three isolates of vLTSV have been sequenced, one each from Australia and New Zealand (KEESE et al. 1983) and one from Canada (ABOUHAIDAR and PALIWAL 1988). Whereas the Australasian isolates differ only in ten of 324 nucleotides, the Canadian isolate of 322 nucleotides has only about 80% sequence homology with them (ABOUHAIDAR and PALIWAL 1988) and should really be considered a separate species. The most highly conserved regions are in the approximately one third of

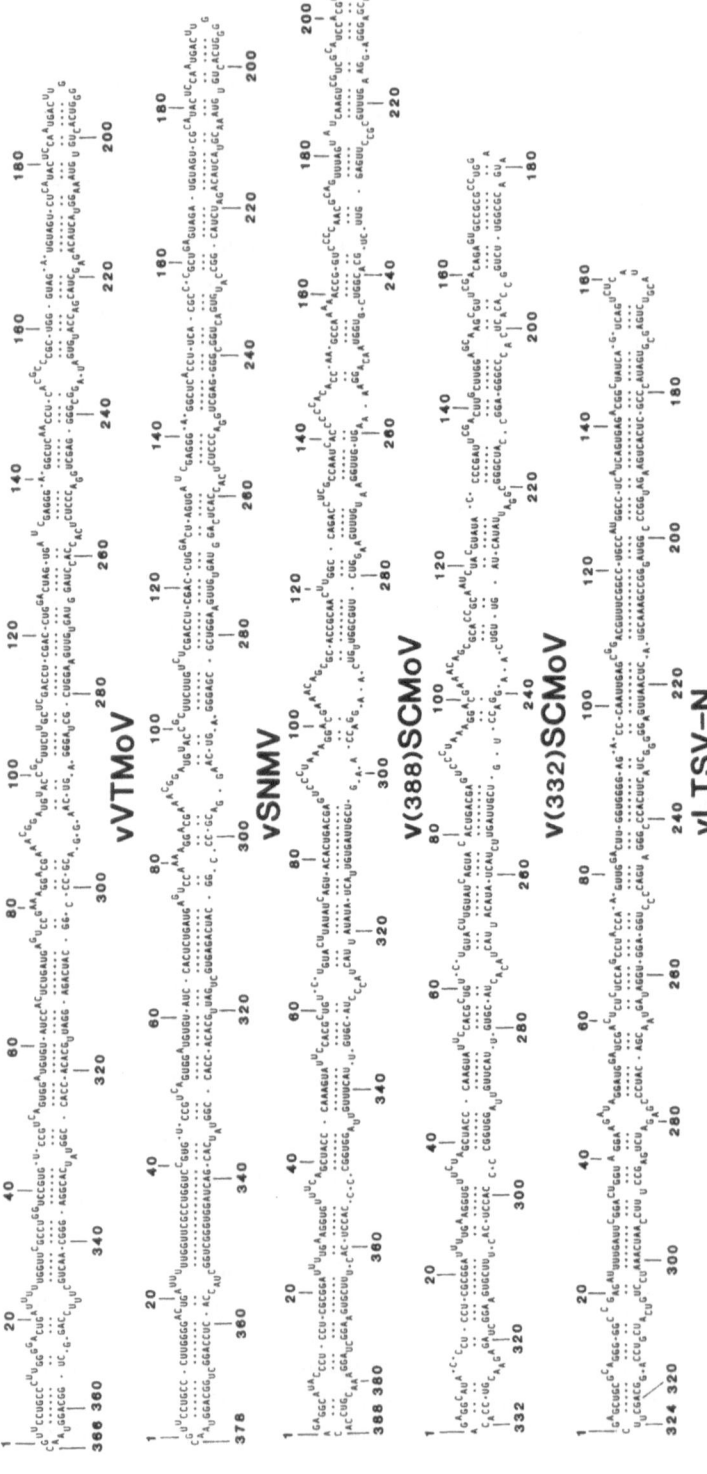

Fig. 4. Sequences and proposed secondary structures of virusoids. Alternative secondary structures can be proposed for various regions of each molecule. (From KEESE and SYMONS 1987, with permission)

the molecule which contains the sequences required for hammerhead self-cleavage (see below). There is very little sequence homology between vLTSV and vVTMoV/vSNMV, except for a few short sequences (KEESE et al. 1983).

In the case of vSCMoV, two species of virusoid were found in different isolates with either one or both species being present (FRANCKI et al. 1983). Sequence analysis (DAVIES et al. 1990) of the 322- and 388-nucleotide species showed a remarkable distribution of homology; the left hand half of v(388)SCMoV (Fig. 4) has a 95% sequence homology with the corresponding part of v(322)SCMoV, whereas the right hand halves have very little sequence homology. The sites of changeover from high to low homology are indicated by filled circles in Fig. 4. Presumably this pattern of homology has arisen through RNA recombination. Such intermolecular rearrangements have been proposed for viroids (KEESE and SYMONS 1985) and other satellite RNAs (SIMON et al. 1988).

At 220nt (COLLINS et al. 1998), vRYMV is very much smaller than the other four virusoids, which vary in length from 322 to 388nt (Table 1). Its predicted native structure is rodlike, with one large branched terminus not predicted in the other virusoids. At the other terminus there is a sequence of 53nt which exhibits 89% sequence identify with vLTSV (COLLINS et al. 1998).

The rodlike secondary structures of virusoids with base-paired regions separated by single-stranded loopouts (Fig. 4) are characteristic of viroids as well as virusoids. The predicted secondary structures were determined by maximizing the number of base pairs, but there are many minor variations possible throughout each molecule. What is most likely of greatest importance is the tertiary structure of each molecule, and this cannot be predicted. As described below (Sect. 7), tertiary structure is central to the highly specific self-cleavage reactions of all virusoids.

Several thermodynamic properties indicate structural differences between viroids and virusoids. Melting profiles of vVTMoV and vSNMV, in which absorbance at 260nm was followed with increasing temperature at defined ionic strength, gave a T_m (temperature at half maximal change in absorbance) that was 10°C lower than that found with viroids, while the breadth of the thermal transition was wider (RANDLES et al. 1982). Further, on the basis of sedimentation coefficients, RIESNER et al. (1982) showed that virusoids were more flexible than viroids, which themselves were more flexible than double-stranded RNA. Overall, therefore, the virusoids investigated have a less stable, more flexible structure in solution than viroids, properties which are determined by both secondary and tertiary interactions.

6 Rolling-Circle Mechanism in the Replication of Virusoids

Circular RNA molecules are considered to replicate via a rolling-circle mechanism (Fig. 5: BRANCH and ROBERTSON 1984; BRANCH et al. 1988). Data in support of this mechanism were originally obtained for viroids, but similar data for virusoids (see below) indicate that they follow a similar pathway.

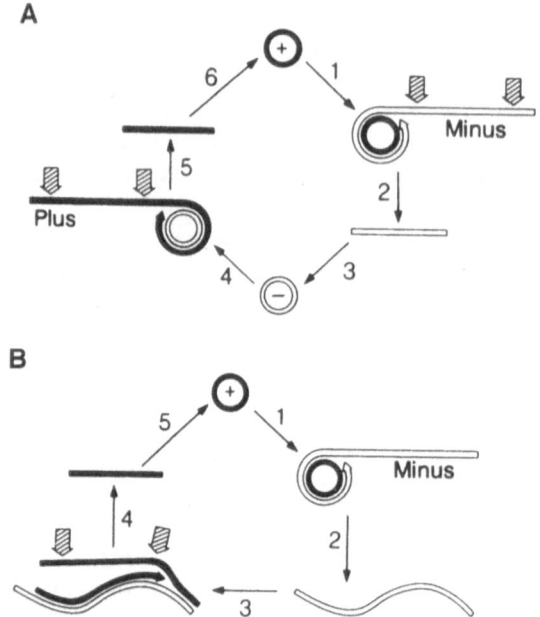

Fig. 5A,B. Two rolling-circle models for the replication of virusoids. See text for details

Two routes for rolling-circle replication are given in Fig. 5. In route (A), a circular-input plus strand is copied by an RNA-dependent RNA polymerase, provided by the host and/or the helper virus, to give an oligomeric minus strand which is cleaved at a specific site to give full-length linear minus strands. These are circularized by a host RNA ligase (BRANCH et al. 1982) and copied by an RNA polymerase, presumably the same one, to give oligomeric plus strands which are specifically cleaved to give monomer plus strands. These are then circularized to give the circular plus progeny virusoids. vLTSV is the only virusoid which appears to follow this route (see below).

In route (B), step 1 is the same; however, the oligomeric minus strand is not cleaved but copied by an RNA polymerase to give an oligomeric plus strand. This is cleaved specifically to give linear plus monomers which are circularized to form the progeny virusoid. All evidence indicates that the other three virusoids follow this route (see below).

Support for the rolling-circle mechanism of replication was provided by Northern hybridization analysis of the different forms of plus and minus virusoid in infected plants (HUTCHINS et al. 1985). The approach used was to prepare partially purified nucleic acid extracts from leaves infected with the helper virus and its virusoid, denature the RNAs by glyoxylation (MCMASTER and CARMICHAEL 1977), and fractionate them by electrophoresis on an agarose gel. Following transfer of the nucleic acids by blotting to nylon membranes, duplicate filters were hybridized with ^{32}P-DNA or RNA probes specific for either plus or minus sequences of each virusoid. The results for one such experiment for vLTSV and vVTMoV are given in Fig. 6.

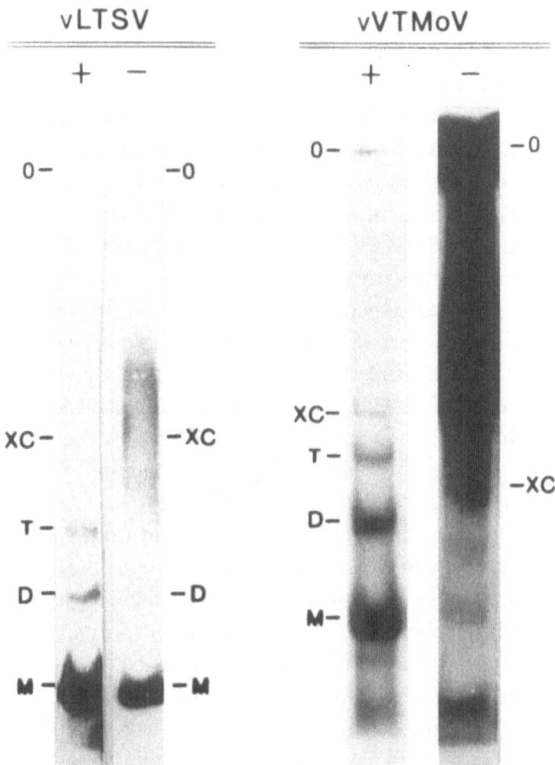

Fig. 6. Northern hybridization analysis of plus and minus sequences of vLTSV and vVTMoV in nucleic acid extracts of infected leaves. Glyoxalated nucleic acid extracts plus marker RNAs were fractionated by electrophoresis in 1.9% agarose gels in 10mM sodium phosphate, pH 6.0. Nucleic acids were bi-directionally transferred to nylon membranes for hybridization with specific plus or minus ^{32}P-cDNA probes. Autoradiography was at room temperature for plus species and at $-70°C$ overnight with an intensifying screen for minus species. (*O* origin of gel, *M*, *D*, *T* monomer, dimer, and trimer species, *XC*, xylene cyanol marker dye.) Full details are given in HUTCHINS et al. (1985)

In the case of vLTSV, plus species consisted mostly of the monomer species (M) with much lower amounts of an oligomeric series (dimer D, trimer T, etc) (Fig. 6). Linear molecules are not separated from circular molecules in the agarose gel electrophoresis system used, but the major monomer species in infected plants is circular (KEESE et al. 1983). The minor amounts of plus oligomers are considered most likely to arise by inefficient cleavage of the long linear plus strand. In the case of minus vLTSV, much lower amounts of minus monomer were found, with no oligomeric series and only a trace of high-molecular-weight material. The similar intensity of the plus and minus vLTSV monomer bands in Fig. 6 is a result of the widely different exposures of the Northern blots; plus species were detected by autoradiography at room temperature and minus species by autoradiography overnight at $-70°C$ with an intensifying screen. These results are consistent with route (A) in Fig. 5.

In the case of vVTMoV (Fig. 6), the plus species were similar to those of vLTSV but there were greater amounts of the oligomers relative to the monomer. In contrast, the minus species consisted of a high-molecular-weight smear, with only trace amounts of a band running in the position of a monomer. Essentially identical results were obtained for vSNMV (HUTCHINS et al. 1985); this is to be expected, in view of the high sequence homology between vVTMoV and vSNMV (HASELOFF and SYMONS 1982). Similar results for vVTMoV have been reported by CHU et al. (1983) and for vSCMoV by DAVIES et al. (1989). These results for these three virusoids are consistent with route (B) of Fig. 5.

Some important practical aspects of these Northern hybridization analyses are worth noting. The minus species of vVTMoV were detected with ^{32}P-RNA but not ^{32}P-DNA probes (HUTCHINS et al. 1985), indicating that they were present in very low amounts. In the case of detection of minus species of vLTSV, when ^{32}P-DNA probes were used, these often gave the same pattern as the plus strands, which indicated cross-hybridization of the minus probe with plus species. However, this problem was not seen using ^{32}P-RNA probes (Fig. 6).

These results emphasize the difficulty in detecting minus species by Northern hybridization analysis when these are present at a much lower level than the plus species. There appear to be two problems. The first was recognized by BRANCH and ROBERTSON (1984), who showed that it is very difficult to detect monomeric minus-sense potato spindle tuber viroid (PSTV) in the presence of excess monomeric plus PSTV. It appears that the excess plus PSTV bound to the nitrocellulose filter was able to hybridize to the bound minus PSTV during the hybridization step and to out-compete the ^{125}I-labeled PSTV that was used as probe. As a consequence, a window of little or no hybridization occurred in the region of the minus species. The extent of this effect is probably highly variable and dependent on the size and nature of the labelled probe as well as the type of membrane used.

The second problem was identified by HUTCHINS et al. (1985), who showed that not all of the RNA transferred to a Genescreen nylon membrane from an agarose gel remained permanently bound after the usual baking at 80°C in vacuo for 2- 4h. Results indicated that greater than 50% of this RNA was released during the prehybridization step, some of which then hybridized to bound complementary sequences. This released and rehybridized RNA could interfere with specific hybridization in three ways:

1. It could prevent hybridization of the probe to filter-bound RNA that was already in double-stranded form. This would make detection of minus species difficult in the presence of excess plus RNA.
2. It could be released into solution and then hybridize with the ^{32}P-probe, thus decreasing its effective concentration.
3. It could form partial hybrids with the ^{32}P-probe while still partially hybridized to filter-bound RNA in a sandwich-type arrangement similar to that described by RANKI et al. (1983). In this way, a ^{32}P-probe for minus sequences would detect bound plus sequences.

In addition to the examples described in HUTCHINS et al. (1985), another example is given by JASPARS et al. (1985) for dot blot hybridization of cucumber mosaic virus plus and minus RNA species. The washing of baked filters at 90°C for 10 min in water prior to hybridization proved to be effective in removing most, but not all, of the loosely bound RNA species (HUTCHINS et al. 1985; JASPARS et al. 1985).

7 Self-cleavage of Virusoid RNAs In Vitro

The specific cleavage of oligomeric plus and minus species produced during the rolling-circle replication of virusoids could occur by the specific action of a host RNase or by an RNA-catalyzed processing or self-cleavage reaction. However, the hammerhead self-cleavage reaction has been demonstrated in vitro for the four virusoids, and this is presumably the reaction that also occurs in vivo (SYMONS 1997).

The self-cleavage reaction in vitro (Fig. 7) occurs in the complete absence of protein and involves Mg^{++} (or other polyvalent cation) catalyzed transphosphorylation, whereby the internucleotide phosphate is transferred from a 3',5'-bond to a 2',3'-cyclic phosphate with the production of a 5'-hydroxyl. The reaction is theoretically reversible.

Self-cleavage in plant pathogenic RNAs in vitro was first described for the linear 359-nucleotide satellite RNA of tobacco ringspot virus (BUZAYAN et al. 1986; PRODY et al. 1986) and for the 247 nucleotide avocado sunblotch viroid (HUTCHINS

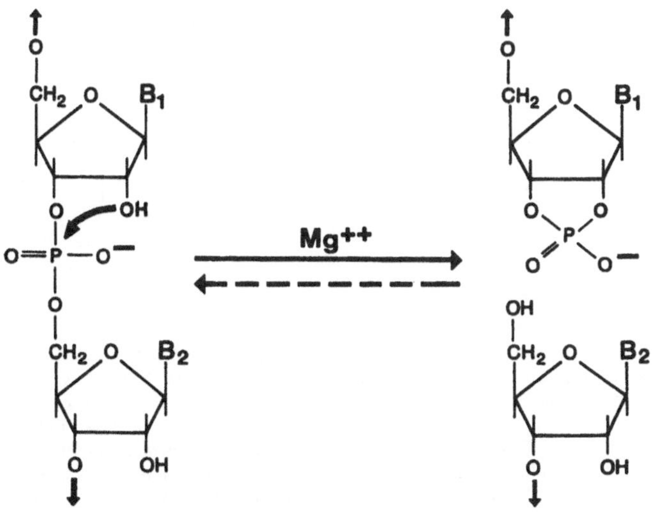

Fig. 7. The self-cleavage reaction of virusoid RNA

et al. 1986; FORSTER et al. 1988). It was also soon demonstrated for vLTSV (FORSTER and SYMONS 1987a,b) and vSCMoV (DAVIES et al. 1989), predicted for vVTMoV and vSNMV (KEESE and SYMONS 1987; FORSTER et al. 1987), and first demonstrated for these two RNAs using short RNA transcripts (McNamara and Symons 1988, unpublished).

The initial approach to investigating self-cleavage in vLTSV was to prepare a partial-length cDNA clone of vLTSV in an SP6 RNA polymerase transcription vector, following a similar route taken to characterize the self-cleavage of avocado sunblotch viroid (HUTCHINS et al. 1986). In this way both plus and minus RNA transcripts were tested for self-cleavage (FORSTER and SYMONS 1987a). Schematic diagrams of plus vLTSV, the plus and minus partial-length cDNA clones, and the products of the RNA transcription reaction are given in Fig. 8A,B.

A

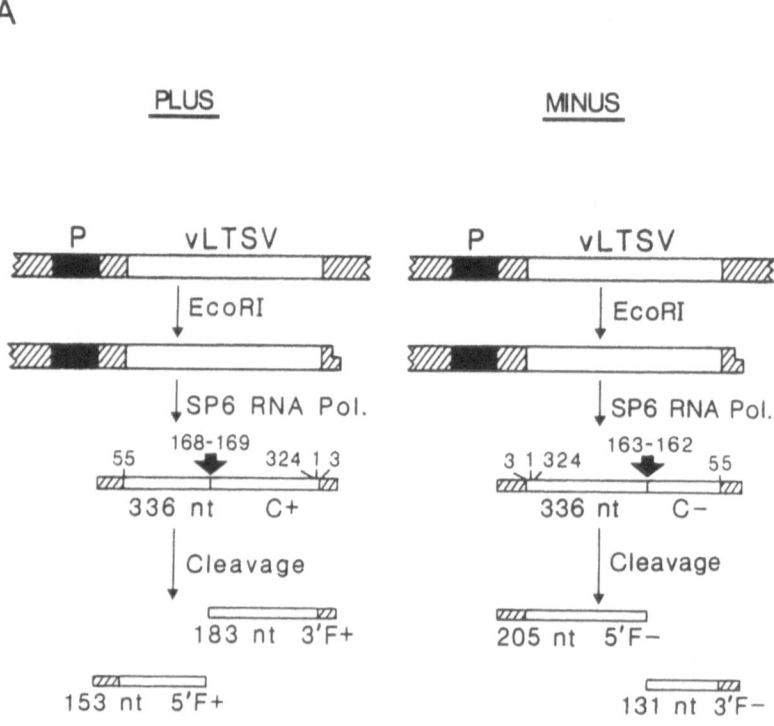

Fig. 8A,B. In vitro synthesis of partial-length plus and minus vLTSV RNAs and self-cleavage reactions of purified RNAs. **A** Diagram of plus and minus DNA templates and SP6 RNA polymerase products generated by transcription of templates linearized with *Eco*RI. Vector sequences are *striped*, the SP6 promoter sequence P is *black*, and vLTSV sequences are *white*. *C+* and *C−*, full-length transcription products; *5'F* and *3'F*, self-cleavage fragments. Nucleotide residue *numbers* of vLTSV given *above* products and length of products *beneath*. Site of cleavage indicated by a *thick arrow*. **B** Analysis of RNAs by electrophoresis on a 5% polyacrylamide, 7*M* urea gel and autoradiography. Products designated on *right hand side* of each gel are as given in **A**, with sizes of single-stranded marker DNAs on the *left hand side*. Further details in text. (Taken from FORSTER and SYMONS 1987a with permission)

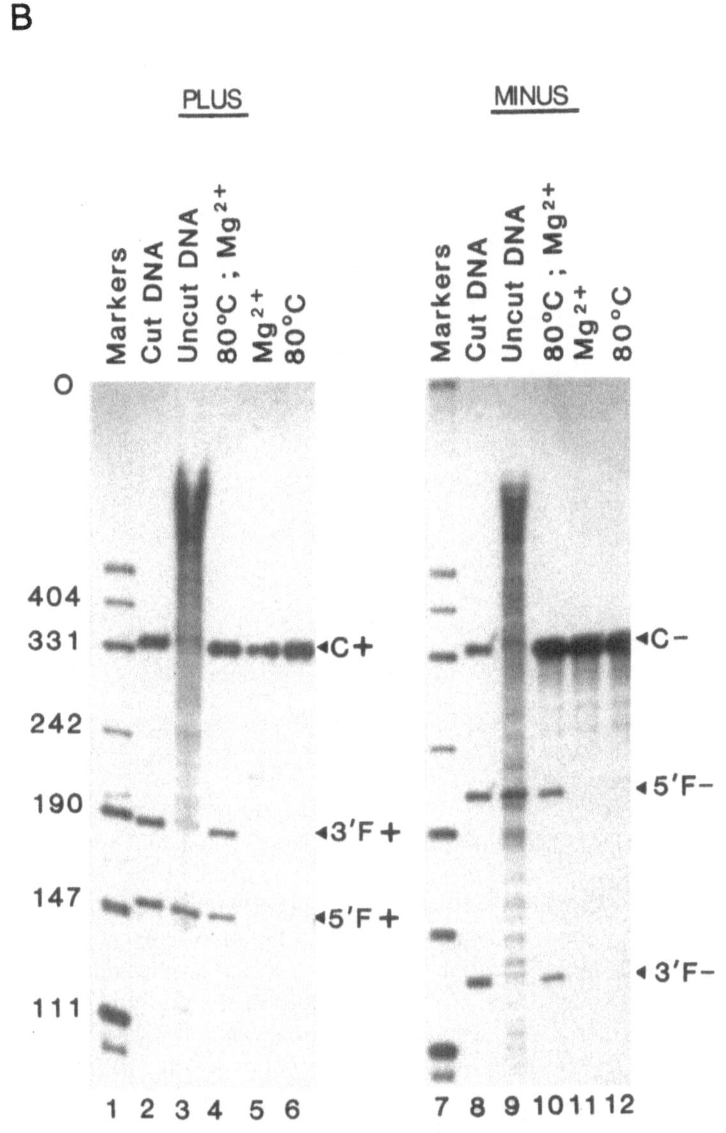

B

Fig. 8A,B. Continued

The plus and minus vLTSV cDNA clones contained residues 55–3 (Fig. 4). When these clones were linearized with *Eco*RI within the polylinker region and transcribed with SP6 RNA polymerase, full-length transcripts of 336 nucleotides were expected (Fig. 8A). In the very first experiment, not only full-length transcripts but also self-cleavage products were obtained. An example of the gel analysis of transcription products prepared under a variety of conditions is given in Fig. 8B, with plus transcripts in lanes 2–6 and minus transcripts in lanes 8–12. Transcripts

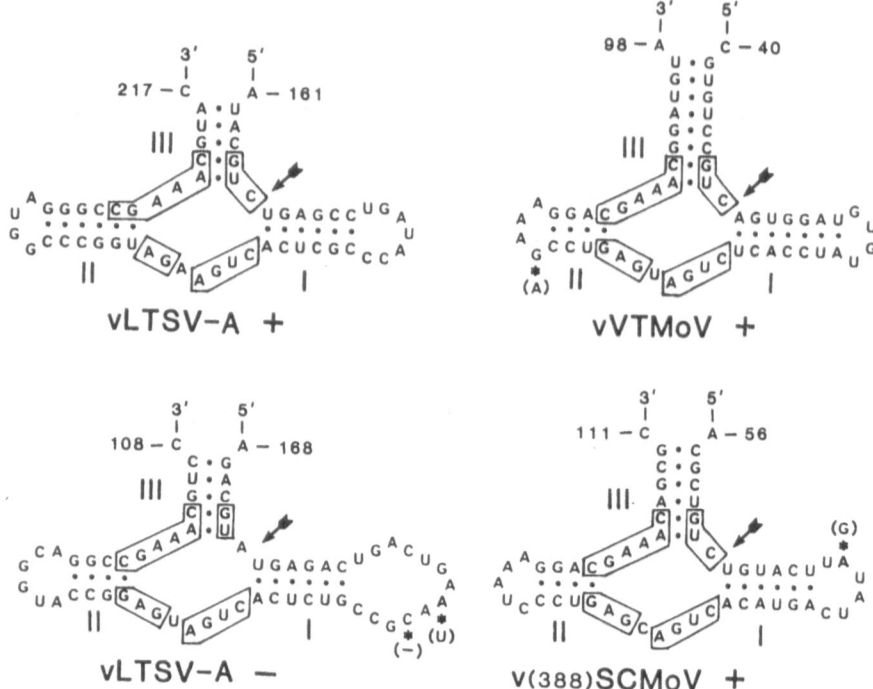

Fig. 9. Proposed hammerhead secondary structures around the self-cleavage site in four virusoid RNAs. The site of cleavage is indicated by an *arrow*. Nucleotides conserved between nine RNAs known to self-cleave via a hammerhead structure are *boxed*

from linearized DNA gave three fragments (lanes 2 and 8), the full-length 336-nucleotide plus and minus transcripts and the two self-cleavage fragments, one derived from the 5'-side of the self-cleavage site (5'F) and the other from the 3'-side (3'F). Uncut plasmid, as predicted, gave only a high-molecular-weight smear plus the 5'F fragment. Characterization of the self-cleavage products provided the self-cleavage sites and showed that the 3'-nucleotides of the plus and minus 5'F fragments terminated in a 2',3'-cyclic phosphate (FORSTER and SYMONS 1987a).

The only protein in these transcription reactions was the SP6 RNA polymerase. To show that the self-cleavage occurred in the complete absence of any protein, the plus and minus 336-nucleotide transcripts were purified from a denaturing polyacrylamide gel and tested for self-cleavage under a range of conditions. For example, incubation of these purified fragments in the presence of 5mM Mg^{++}, pH 7.5, for 10 min at 25°C gave no self-cleavage (Fig. 8B, lanes 5 and 11). However, if the purified RNAs were heated in 1mM EDTA for 1min at 80°C, followed by snap-cooling on ice, prior to the addition of Mg^{++} at 0°C and incubation at 25°C, then significant self-cleavage occurred (lanes 4 and 10). The other control of preheating/snap-cooling and incubation at 25°C in the absence of Mg^{++} gave no self-cleavage (lanes 6 and 12). Hence, the gel-purified transcripts were inactive in self-cleavage unless heated and snap-cooled and then incubated with Mg^{++}, a

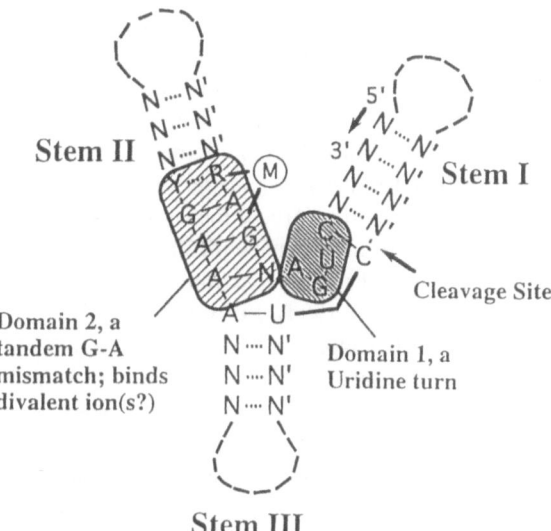

Fig. 10. Schematic drawing of the structure of the hammerhead as determined by X-ray crystallography. (Reproduced from McKay 1996 with permission)

result which indicates that the gel purification procedure provides RNA in an inactive conformation, some of which can be converted into an active self-cleavage structure by heating and snap-cooling.

These results with plus and minus vLTSV RNA were identical to those previously reported for self-cleavage of a dimeric transcript of avocado sunblotch viroid (Hutchins et al. 1986) and of transcripts of the plus satellite RNA of tobacco necrosis virus (Buzayan et al. 1986), in that self-cleavage is a simple transphosphorylation reaction, as seen in Fig. 7. The same reaction has been shown to occur in vSCMoV, but only for the plus RNA; no self-cleavage occurs in the minus RNA (Davies et al. 1989).

8 Hammerhead-shaped Secondary Structure Model for the Self-cleavage Site

The proposed secondary structures around the demonstrated self-cleavage sites of plus and minus vLTSV (Forster and Symons 1987a) and plus vSCMoV (Davies et al. 1989) and the putative self-cleavage sites of plus vVTMoV and vSNMV (Keese and Symons 1987; Symons et al. 1987) are shown in Fig. 9. These hammerhead structures all contain three base-paired stems (I, II, and III) formed around a single-stranded interior loop with the self-cleavage sites located in identical positions. There are 13 nucleotides (boxed) which are conserved among the nine RNAs identified so far where self-cleavage is mediated via a hammerhead structure; in addition to the five RNAs in Fig. 7, there are the plus and minus RNAs of avocado sunblotch viroid (HUTCHINS et al. 1986), the plus RNA of satellite tobacco ringspot virus (BUZAYAN et al. 1986), and the transcript of the 330-bp satellite DNA of the newt (EPSTEIN and GALL 1987).

The simple two-dimensional model of the hammerhead structure (Fig. 9) has been very useful since 1986 for designing experiments to characterize the self-cleavage reaction. However, it is the three-dimensional structure of the hammerhead which is required to determine its mechanism of action. Nine years after the initial publications of the self-cleavage reaction, the first analysis of a crystal structure was published (PLEY et al. 1994). For this work, a 34-mer ribozyme and a 13-mer noncleavable deoxynucleotide substrate were hybridized together and crystallized and the X-ray structure was determined to 2.6Å. Soon thereafter, the structure for an all-RNA system was reported and self-cleavage was blocked by a 2'-0-methyl group on the 2'-hydroxyl at the cleavage site (SCOTT et al. 1995). More recently, the crystal structure of an unmodified hammerhead RNA sequence was determined in the absence of a divalent metal ion (SCOTT et al. 1996). Self-cleavage occurred in the crystal on the addition of divalent metal ions.

It is of interest that the three-dimensional structures determined from the three approaches are essentially the same. A diagrammatic representation of the structure is given in Fig. 10 (McKAY 1996). Stems I and II are not linear as in the two-dimensional model but form the arms of a Y-shaped structure, and stem III now forms its stem.

9 Fraction of Virusoid Molecule involved in Self-cleavage

Given that self-cleavage is a key reaction in the rolling-circle replication of all virusoids in vivo, this is the only function that can be ascribed so far to virusoid RNA. Although virusoids have both AUG- and GUG-initiated open reading frames, there is no evidence to indicate that these code for functional proteins in vivo (HASELOFF and SYMONS 1982; KEESE and SYMONS 1987). In the case of vLTSV, the plus and minus self-cleavage sites (Fig. 11A) are only six residues apart at the right hand end of the molecule (Fig. 11B). The predicted hammerhead structures of plus and minus vLTSV have 43 conserved nucleotides (boxed in Fig. 11A); this is somewhat surprising, since the sequences are derived from complementary RNA strands and they partially overlap (Fig. 11B). However, most of the sequences do come from that part of the vLTSV molecule where there is extensive base pairing.

The total sequences which make up the plus and minus self-cleavage structures are about one third of the vLTSV molecule. It is possible that self-cleavage may be the main or the only function of this region of the molecule. The single plus self-cleavage site of each of the other three virusoids occurs in the top strand of the left hand half of each molecule and, unlike vVLTSV, the sequences are not derived

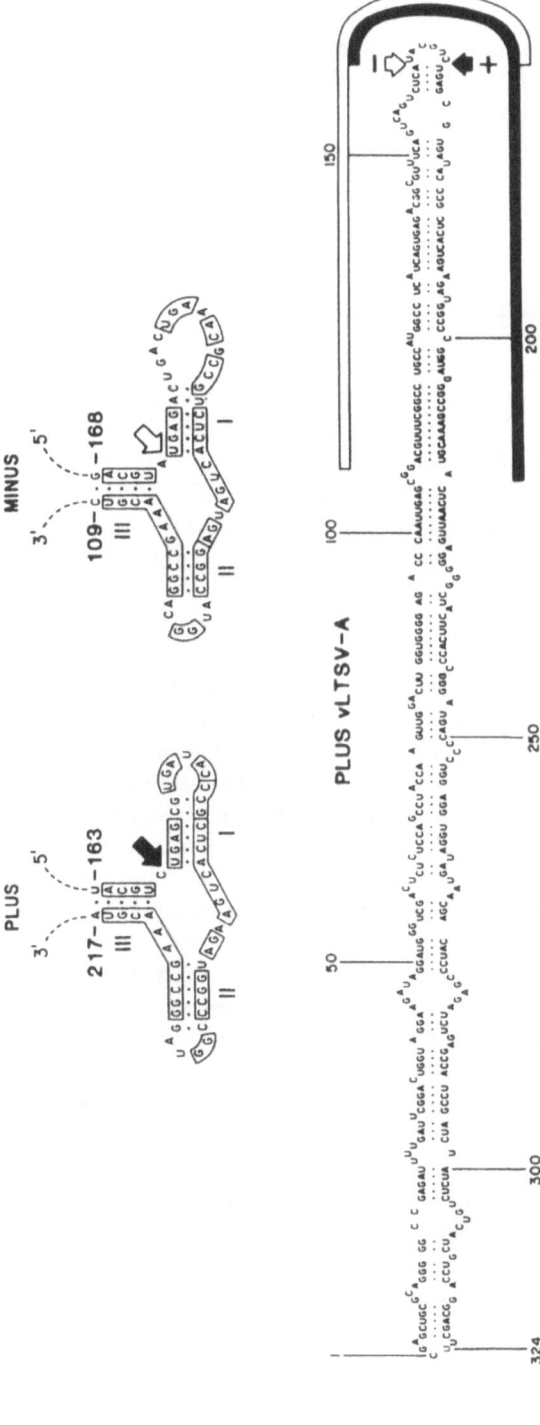

Fig. 11A,B. Location of self-cleavage domains for plus and minus hammerhead structures of vLTSV. **A** Extensive sequence homology between the plus and minus hammerhead structures. Conserved nucleotides are boxed and self-cleavage sites indicated by an *arrow*. **B** Self-cleavage domains within the proposed native secondary structure of vLTSV. *Arrows* show the plus and minus self-cleavage sites. *Black and white bars* indicate the sequences that constitute the plus and minus hammerhead structures, respectively, in A. (Taken from FORSTER and SYMONS 1987a, with permission)

from parts of the molecule which are partially double stranded. The 50–60 nucleotides which make up the predicted hammerhead self-cleavage structures of vVTMoV, vSNMV, and vSCMoV (Fig. 9) account for 15–20% of each molecule; taken together with the opposite, partially base-paired strand of each virusoid, the total accounts for about one third of each molecule, a proportion similar to that of vLTSV.

10 Circularization of Linear Monomers Produced by Self-cleavage

Since virusoids exist as covalently closed circular monomers in infected cells (FRANCKI 1987), the linear monomers produced during rolling-circle replication must be circularized. RNA ligases exist in plants, e.g., wheat germ and *Chlamydomonas*, which specifically require 2′,3′-cyclic phosphate and 5′-phosphorylated termini for ligation (KONARSKA et al. 1982; KIKUCHI et al. 1982). A 2′-phosphate occurs at the ligation site in addition to the normal 3′-5′-phosphodiester linkage.

KIBERSTIS et al. (1985) identified a 2′-phosphate on the C residue at the predicted self-cleavage site of circular vSNMV and vVTMoV purified from infected plants; this is consistent with the mechanism of action of a plant RNA ligase and the rolling-circle mechanism of replication. Monomeric, linear avocado sunblotch viroid (ASBV) produced by self-cleavage of plus and minus dimeric transcripts was efficiently ligated to monomer circles in vitro using partially purified wheat germ RNA ligase (HUTCHINS et al. 1986). However, a 2′-phosphate was not found at the self-cleavage site of circular ASBV purified from infected avocado leaves (Rathjen and Symons, unpublished), even though the mechanism of production and circularization of linear monomers in vivo is predicted to be similar (HUTCHINS et al. 1986). An explanation consistent with these two sets of results is that the 2′-phosphates produced in the circularization of ASBV are removed by a plant phosphatase. On the other hand, the 2′-phosphates of vVTMoV may be protected from such phosphatase action by rapid encapsidation in progeny virions, as suggested by KIBERSTIS et al. (1985).

An alternative explanation needs to be considered for ASBV, which is found only in the chloroplasts of infected avocado leaves, as determined by in situ hybridization (BONFIGLIOLI et al. 1996), and presumably also replicates there. Since chloroplasts are bacterium-like organelles, the RNA ligase involved in the circularization of linear ASBV could have properties similar to those of the bacteriophage T4 RNA ligase, which requires a 5′-phosphate and 3′-hydroxyl group for circularization, and there is no formation of a 2′-O-phosphate (UHLENBECK and GUMPORT 1982).

11 Information Content of Virusoids

Limited indirect evidence indicates that virusoids do not code for proteins in vivo. vVTMoV, vSNMV, and vLTSV contain three to six AUG-initiated open reading frames in their plus and minus strands. However, there is no conservation of potential translation products between vSNMV and the 365- and 366-nucleotide forms of vVTMoV, despite their approximately 94% sequence homology (HASELOFF and SYMONS 1982). In the case of the two closely related sequence variants of vLTSV isolated in Australia and New Zealand, only two of the seven potential AUG-initiated polypeptides are conserved in terms of the sites of their start and stop codons and substantial sequence homology (KEESE et al. 1983). Only one peptide of 25 amino acids in the Canadian isolate of vLTSV is similar to putative peptides of approximately the same size in the two Australasian isolates (ABOU-HAIDAR and PALIWAL 1988). Furthermore, no in vitro translation products have been attributed to vLTSV (MORRIS-KRSINICH and FORSTER 1983), while eukaryotic ribosomes do not bind to circular RNA, an essential step for the initiation of translation (KOZAK 1979).

The sequences involved in the self-cleavage reaction in vitro provide the only evidence so far for in vivo function of any part of the virusoid molecule. Approximately one third of each molecule appears to be involved here. The role of the self-cleavage reaction characterized in vitro in the rolling-circle replication of the virusoids in vivo is inferred from the putative intermediates isolated from infected plants (HUTCHINS et al. 1985), as considered above. In order to provide more definitive evidence, three full-length cDNA clones of vLTSV were mutated at different sites, such that in vitro self-cleavage in the minus strands was eliminated (SHELDON and SYMONS 1993). It was predicted that such mutations would lead to the accumulation of multimeric minus vLTSV strands in vitro.

When the excised inserts from these mutated cDNA clones were co-inoculated with the helper virus LTSV on an herbaceous host plant, high-molecular-weight minus vLTSV was detected as expected, while the wild-type control vLTSV gave only the unit length minus vLTSV. However, it was a surprise to find that the mutated virusoids also produced some monomeric minus species, as shown by Northern hybridization analysis. Sequence analysis of the plus monomeric progeny of the three mutated cDNA clones showed that 8–20% of the progeny contained reversions and pseudoreversions of the introduced mutations which allowed self-cleavage to occur in the minus strand. Overall, the results provide convincing evidence for the predicted role of hammerhead self-cleavage in vLTSV replication.

There must be promoter sequences in both plus and minus RNAs for the initiation of complementary RNA synthesis by an RNA-dependent RNA polymerase which is presumably provided by the helper virus. In the absence of definitive evidence so far, it must also be considered that the helper virus may provide a protein cofactor(s) that allows modification of a host enzyme to carry out the replication function; this is not that unlikely, as viroids must be replicated by a host enzyme since a helper virus is not needed.

Virusoids are readily encapsidated in the virions of the helper virus; recognition signals for packaging must therefore be encoded in the virusoids. As regards other potential functions of the virusoid RNAs, we can only speculate. For example, what is the nature of the reasonably specific interaction between the helper virus RNA and the virusoid? Is there some type of antisense interaction, as has been reported for the 336-nucleotide linear satellite RNA of cucumber mosaic virus where a stable hybrid can be formed in vitro between the satellite RNA and the coat protein gene of the viral RNA (REZAIAN and SYMONS 1986)? However, whether this in vitro observation has any relevance in vivo has not been determined. What is the molecular basis by which a virusoid can modify the symptom expression of its helper virus? In the case of LTSV, which produces very mild chlorotic lesions on *Chenopodium quinoa*, necrotic lesions are produced in the presence of vLTSV (FRANCKI 1987). Does a virusoid interact with specific host proteins with recognition via its secondary and tertiary structure? Can a virusoid interfere with host RNA metabolism by competing with host RNA:RNA interactions or even by forming antisense complexes.

An intriguing aspect of 12 viroids examined, including ASBV and peach latent mosaic viroid, as well as the 1700-nt circular hepatitis delta RNA, which has viroid-like features (BRANCH et al. 1990), is the pattern of polypurine and polypyrimidine tracts which make up substantial portions of their RNAs (BRANCH et al. 1993). The biological significance of these tracts is not known, but they are not present in the four virusoids and sTRSV. However, it was suggested that the presence or absence of these tracts may be connected in some way with the differences in replication strategies of the two groups of RNAs, with viroid replication being independent of a helper virus whereas virusoid and sTRSV replication is dependent on a helper virus (BRANCH et al. 1993).

Obviously, there are many questions to be answered. The small size of virusoids probably belies the large amount of genetic information that is elegantly packed into these molecules and remains to be deciphered.

References

Abouhaidar MG, Paliwal YC (1988) Comparison of the nucleotide sequences of viroid-like satellite RNA of the Canadian and Australasian strains of lucerne transient streak virus. J Gen Virol 69:2369 2373
Blackstock JMcK (1978) Lucerne transient streak and lucerne latent, two new viruses of lucerne. Aust J Agric Res 29:291 304
Bonfiglioli RG, Webb DR, Symons RH (1996) Tissue and intracellular distribution of coconut cadang cadang viroid and citrus exocortis viroid determined by in situ hybridisation and confocal laser scanning and transmission electron microscopy. Plant J 9:457 465
Branch AD, Robertson HD (1984) A replication cycle for viroids and other small infectious RNAs. Science 223:450 455
Branch AD, Robertson HD, Greer C, Gegenheimer P, Peebles C, Abelson J (1982) Cell-free circularization of viroid progeny RNA by an RNA ligase from wheat germ. Science 217:1147 1149
Branch AD, Benenfeld BJ, Robertson HD (1988) Evidence for a single rolling circle in the replication of potato spindle tuber viroid. Proc Natl Acad Sci USA 85:9128 9132

Branch AD, Levine BJ, Robertson HD (1990) The brotherhood of circular RNA pathogens: viroids, circular satellites, and the delta agent. Semin Virol 1:143–152

Branch AD, Lee SE, Neel OD, Robertson HD (1993) Prominent polypurine and polypyrimidine tracts in plant viroids and in RNA of the human hepatitis delta virus. Nucleic Acids Res 21:3529–3535

Buzayan JM, Gerlach WL, Bruening G (1986) Non-enzymatic cleavage and ligation of RNAs complementary to a plant virus satellite RNA. Nature 323:349–353

Chu PWG, Francki RIB, Randles JW (1983) Detection, isolation and characterization of high molecular weight double-stranded RNAs in plants infected with velvet tobacco mottle virus. Virology 126: 480–492

Collins RF, Gellatly DL, Sehgal OP, Abouhaidar MG (1998) Self-cleaving RNA associated with rice yellow mottle virus is the smallest viroid-like RNA. Virology 241:269–275

Dall DJ, Graddon DJ, Randles JW, Francki RIB (1990) Isolation of a subterranean clover mottle virus-like satellite RNA from lucerne infected with lucerne transient streak virus. J Gen Virol 71:1873–1875

Davies C, Haseloff J, Symons RH (1990) Structure, self-cleavage and replication of two viroid-like circular satellite RNAs (virusoids) of subterranean clover mottle virus. Virology 177:216–224

Epstein LM, Gall JG (1987) Self-cleaving transcripts of satellite DNA from the newt. Cell 48:535–543

Forster RLS, Jones AT (1979) Properties of lucerne transient streak virus, and evidence of its affinity to southern bean mosaic virus. Ann Appl Biol 93:181–189

Forster AC, Symons RH (1987a) Self-cleavage of plus and minus RNAs of a virusoid and a structural model for the active sites. Cell 49:211–220

Forster AC, Symons RH (1987b) Self-cleavage of virusoid RNA is performed by the proposed 55-nucleotide active site. Cell 50:9–16

Forster AC, Jefferies AC, Sheldon CC, Symons RH (1987) Structural and ionic requirements for self-cleavage of virusoid RNAs and trans self-cleavage of viroid RNA. Cold Spring Harbor Symp Quant Biol 52:249–259

Forster AC, Davies C, Sheldon CC, Jeffries AC, Symons RH (1988) Self-cleaving viroid and newt RNAs may only be active as dimers. Nature 334:265–267

Francki RIB (1985) Plant virus satellites. Annu Rev Microbiol 39:151–174

Francki RIB (1987) Possible viroid origin. Encapsidated viroid-like RNA. In: Diener TO (ed) The viroids. Plenum, New York, pp 205–218

Francki RIB, Randles JW, Hatta T, Davies C, Chu PWG, McLean GD (1983) Subterranean clover mottle virus: another virus from Australia with encapsidated viroid-like RNA. Plant Pathol 32:47–59

Francki RIB, Zaitlin M, Palukaitis P (1986a) In vivo encapsidation of potato spindle tuber viroid by velevet tobacco mottle virus particles. Virology 155:469–473

Francki RIB, Grivell CJ, Gibb KS (1986b) Isolation of velvet tobacco mottle virus capable of replication with and without a viroid-like RNA. Virology 148:381–384

Gould AR (1981) Studies on encapsidated viroid-like RNA. II. Purification and characterization of a viroid-like RNA associated with velvet tobacco mottle virus (VTMoV). Virology 108:123–133

Gould AR, Hatta T (1981) Studies on encapsidated viroid-like RNA. III. Comparative studies on RNAs isolated from velvet tobacco mottle virus and Solanum nodiflorum mottle virus. Virology 109:137–147

Gould AR, Francki RIB, Randles JW (1981) Studies on encapsidated viroid-like RNA. IV. Requirement for infectivity and specificity of two RNA components from velvet tobacco mottle virus. Virology 110:420–426

Greber RS (1973) A beetle-transmitted isometric virus from Solanum nodiflorun. Aust Plant Pathol Newslett 2:3

Greber RS (1981) Some characteristics of Solanum nodiflorum mottle virus – a beetle-transmitted isometric virus from Australia. Aust J Biol Sci 34:369–378

Greber RS, Randles JW (1986) Solanum nodiflorum mottle virus. AAB descriptions of plant viruses, no 318. Association of Applied Biologists, Wellesbourne, UK

Haseloff J, Symons RH (1982) Comparative sequence and structure of viroid-like RNAs of two plant viruses. Nucleic Acids Res 10:368–369

Haseloff J, Mohamed NA, Symons RH (1982) Viroid RNAs of the cadang-cadang disease of coconuts. Nature 229:316–321

Hutchins CJ, Keese P, Visvader JE, Rathjen PD, McInnes JL, Symons RH (1985) Comparison of multimeric plus and minus forms of viroids and virusoids. Plant Mol Biol 4:293–304

Hutchins CJ, Rathjen PD, Forster AC, Symons RH (1986) Self-cleavage of plus and minus RNA transcripts of avocado sunblotch viroid. Nucleic Acids Res 14:3627–3640

Jaspars EMJ, Gill DS, Symons RH (1985) Viral RNA synthesis by a particulate fraction from cucumber seedlings infected with cucumber mosaic virus. Virology 144:410–425

Jones AT, Mayo MA (1983) Interaction of lucerne transient streak virus and the viroid-like RNA-2 of Solanum nodiflorum mottle virus. J Gen Virol 64:1771–1774

Jones AT, Mayo MA (1984) Satellite nature of the viroid-like RNA-2 of Solanum nodiflorum mottle virus and the ability of other plant viruses to support the replication of viroid-like RNA molecules. J Gen Virol 65:1713–1721

Jones AT, Mayo MA, Duncan GH (1983) Satellite-like properties of small circular RNA molecules in particles of lucerne transient streak virus. J Gen Virol 64:1167–1173

Keese P, Symons RH (1985) Domains in viroids: evidence of intermolecular RNA rearrangements and their contribution to viroid evolution. Proc Natl Acad Sci USA 82:4582–4586

Keese P, Symons RH (1987) The structure of viroids and virusoids. In: Semancik JS (ed) Viroids and viroid-like pathogens. CRC Press, Boca Raton, pp 1–47

Keese P, Bruening G, Symons RH (1983) Comparative sequence and structure of circular RNAs from two isolates of lucerne transient streak virus. FEBS Lett 159:185–190

Kiberstis PA, Haseloff J, Zimmern D (1985) 2′ Phosphomonoester, 3′-5′ phosphodiester bond at a unique site in a circular viral RNA. EMBO J 4:817–827

Kikuchi Y, Tyc K, Filipowicz W, Sänger HL, Gross HJ (1982) Circularization of linear viroid RNA via 2′-phosphomonoester, 3′, 5′-phosphodiester bonds by a novel type of RNA ligase from wheat germ and Chlamydomonas. Nucleic Acids Res 10:7521–7529

Konarska M, Filipowicz W, Gross HJ (1982) RNA ligation via 2′-phosphomonoester, 3′,5′-cyclic phosphate termini and involvement of a 5′-hydroxyl polynucleotide kinase. Proc Natl Acad Sci USA 79:1474–1478

Kozak M (1979) Inability of circular mRNA to attach to eukaryotic ribosomes. Nature 280:82–85

Mäkinen K, Tamm T, Nss V, Truve E, Puurand Ü, Munthe T, Saarma M (1995) Characterization of cocksfoot mottle sobemovirus genomic RNA and sequence comparison with related viruses. J Gen Virol 76:2817–2825

McKay DB (1996) Three-dimensional structure of the hammerhead ribozyme. Nucleic Acids Mol Biol 10:161–172

McMaster GK, Carmichael GG (1977) Analysis of single- and double-stranded nucleic acids on polyacrylamide gels by using glyoxal and acridine orange. Proc Natl Acad Sci USA 74:4835–4838

Milligan JF, Groebe DR, Witherell GW, Uhlenbeck OC (1987) Oligoribonucleotide synthesis using T7 RNA polymerase and synthetic DNA templates. Nucleic Acids Res 15:8783–8798

Morris-Krsinich BAM, Forster RLS (1983) Lucerne transient streak virus RNA and its translation in rabbit reticulocyte lysate and wheat germ extract. Virology 128:176–185

Murant AF, Mayo MA (1982) Satellites of plant viruses. Annu Rev Phytopathol 20:49–70

Othman Y, Hull R (1995) Nucleotide sequence of the bean strain of southern bean mosaic virus. Virology 206:287–297

Paliwal YC (1983) Identification and distribution in eastern Canada of lucerne transient streak, a virus newly recognised in North America. Can J Plant Pathol 5:75–80

Paliwal YC (1984) Interaction of the viroid-like RNA-2 of lucerne transient streak virus with southern bean mosaic virus. Can J Plant Pathol 6:93–97

Pley HW, Flaherty KM, McKay DB (1994) Three-dimensional structure of a hammerhead ribozyme. Nature 372:68–74

Prody GA, Bakos JT, Buzayan JM, Schneider IR, Bruening G (1986) Autolytic processing of dimeric plant virus satellite RNA. Science 231:1577–1580

Randles JW, Francki RIB (1986) Velevet tobacco mottle virus. AAB description of plant viruses, no 317. Association of Applied Biologists, Wellesbourne, UK

Randles JW, Davies C, Hatta T, Gould AR, Francki RIB (1981) Studies on encapsidated viroid-like RNA. I. Characterization of velvet tobacco mottle virus. Virology 108:111–122

Randles JW, Steger G, Riesner D (1982) Structural transitions in viroid-like RNAs associated with cadang-cadang disease, velvet tobacco mottle virus and Solanum nodiflorum mottle virus. Nucleic Acids Res 10:5569–5586

Ranki M, Palva A, Virtanen M, Laaksonen M, Söderlund H (1983) Sandwich hybridization as a convenient method for the detection of nucleic acids in crude samples. Gene 21:77–85

Rezaian MA, Symons RH (1986) Anti-sense regions in satellite RNA of cucumber mosaic virus from stable complexes with the viral coat protein gene. Nucleic Acids Res 14:3229–3239

Riesner D, Kaper JM, Randles JW (1982) Stiffness of viroids and viroid-like RNA in solution. Nucleic Acids Res 10:5587–5598

Scott WG, Finch JT, Klug A (1995) The crystal structure of an all-RNA hammerhead ribozyme: a proposed mechanism for RNA catalytic cleavage. Cell 81:991–1002

Scott WG, Murray JB, Arnold JRP, Stoddard BL, Klug A (1996) Capturing the structure of a catalytic RNA intermediate: the hammerhead ribozyme. Science 274:2065–2069

Sehgal OP, Sinha RC, Gellatly DL, Ivanov I, AbouHaidar MG (1993) Replication and encapsidation of the viroid-like satellite RNA of lucerne transient streak virus are supported in divergent hosts by cocksfoot mottle virus and turnip rosette virus. J Gen Virol 74:785–788

Sheldon CC, Symons RH (1989) Mutagenesis analysis of a self-cleaving RNA. Nucleic Acids Res 17:5679–5685

Simon AE, Engel H, Johnson RP, Howell SH (1988) Identification of regions affecting virulence, RNA processing and infectivity in the virulent satellite of turnip crinkle virus. EMBO J 7:2645–2651

Symons RH (1997) Plant pathogenic RNAs and RNA catalysis. Nucleic Acids Res 25:2683–2689

Symons RH, Hutchins CJ, Forster AC, Rathjen PD, Keese P, Visvader JE (1987) Self-cleavage of RNA in the replication of viroids and virusoids. J Cell Sci [Suppl] 7:303–318

Tien P, Davies C, Hatta T, Francki RIB (1981) Viroid-like RNA encapsidated in lucerne transient streak virus. FEBS Lett 132:353–356

Uhlenbeck OG, Gumport RI (1982) T4 RNA ligase. In the Enzymes 15:31–60. Academic Press, New York

Wu S, Rinehart CA, Kaesberg P (1987) Sequence and organisation of southern bean mosaic virus genomic RNA. Virology 16:73–80

Yassi MNA, Ritzenthaler C, Brugidou C, Fauquet CM, Beachy RN (1994) Nucleotide sequence and genome characterization of rice yellow mottle virus RNA. J Gen Virol 75:249–257

Human Hepatitis Delta Virus: an Agent with Similarities to Certain Satellite RNAs of Plants

J.M. TAYLOR

Fox Chase Cancer Center, 7701 Burholme Avenue, Philadelphia, PA 19111, USA
J.T. was supported by grants AI-26522 and CA-06927 from the N.I.H. and by an appropriation from the
Commonwealth of Pennsylvania

1 Introduction

1.1 Objectives

One of the most intriguing aspects of hepatitis delta virus (HDV) is that, even though it is a naturally occurring infectious agent of human beings, it nevertheless shows a number of remarkable similarities to viroids, the subviral agents of plants that are considered elsewhere in this volume. The recognition and exploitation of this analogy has been a major contributor to studies of HDV genome replication. This review will concentrate on the structure of the HDV genome and the mechanism by which the genome is replicated, with special emphasis on similarities to the viroids. The review will not consider in any depth other important aspects of HDV, such as the assembly of genomes into virus particles, experimental transmission, host responses, or epidemiology. For such information the reader is directed to other more extensive reviews (LAI 1995; PURCELL and GERIN 1996; HADZIYANNIS 1997; MARGOLIS et al. 1997). However, to put this review in perspective, we will present a limited amount of background material.

1.2 Background Information

1.2.1 Helper Virus

HDV was discovered 20 years ago by RIZZETTO et al. in chronic hepatitis B virus carriers experiencing expanded attacks of acute liver disease. The discovery was facilitated by the fact that the virus expresses a protein, referred to as the delta antigen (δAg) which gave rise to a strong humoral immune response in these patients (RIZZETTO et al. 1977). In 1986 it was discovered that HDV has a single-stranded RNA genome that is actually circular in conformation (WANG et al. 1986). This finding rapidly led to a series of studies which showed extensive similarities between the structure and replication of the HDV genome and some of the plant viroids (TAYLOR et al. 1990).

In nature, HDV is always found in human beings in association with infections by hepatitis B virus (HBV). The HDV particles present in the serum contain the RNA genome and the δAg in an RNP complex that is surrounded by a lipid-containing envelope. This envelope is produced by the so-called surface antigen proteins of the helper virus, HBV. Since a full cycle of HDV replication is dependent upon the utilization of these HBV proteins, HDV does not satisfy the definition of a virus; it is a subviral satellite which uses HBV as its helper (DIENER and PRUSINER 1985).

Typical HBV infections of the liver produce up to 10^9 infectious particles per milliliter of serum. This virus is usually associated with a 100-fold or greater excess of smaller empty particles, referred to as surface antigen particles, which are synthesized because the HBV envelope proteins are 'over produced'. In the liver of a patient infected with both HBV and HDV, the HDV RNA genome, along with δAg, is packaged by utilization of the excess surface proteins produced by HBV. The titer of such HDV particles can be as high as 10^{12} particles per milliliter (PONZETTO et al. 1991). Since HBV and HDV particles share the same envelope proteins, it is assumed that both viruses share the same host cellular receptors for infection, even though such receptors are as yet not specifically identified.

In nature, HDV replicates only in human hosts. Experimentally, it can be transmitted to chimpanzees (RIZZETTO et al. 1980). It is also possible to replace the HBV helper virus with a related virus, woodchuck hepatitis virus (WHV), and to establish HDV replication in woodchucks (PONZETTO et al. 1984).

1.2.2 Infectious Clones

It has been possible to obtain cDNA clones of the small 1700-nt HDV RNA genome and thus express a variety of permuted monomers and multimers of HDV RNA (KUO et al. 1989). HDV genome replication can be initiated in mammalian cells transfected with either RNA or cDNA (KUO et al. 1989; GLENN et al. 1990). An important qualification to this general statement is that transfected RNA alone does not initiate genome replication. It does if the cell is already expressing the small delta antigen (δAg-S)(see Sect. 1.2.4). The role(s) of δAg-S in this transfection is not clear. Possibilities include stabilization of transfected RNA and/or facilitated transport of transfected RNA to the nucleus following δAg-S binding.

We think that genome replication initiated by such transfections differs from the natural total replication cycle in human beings only in that the envelope proteins of the helper virus are absent. Thus, there is no assembly and release of infectious RNA-containing particles.

1.2.3 Three Stable RNAs

Figure 1 shows the three RNAs that are associated with the replication of HDV. The first species is the genome, which by definition is the RNA that is normally found inside virus particles. Like many of the viroid RNAs, it is single stranded, circular, and predicted to fold into an unbranched rodlike structure with about 70% of the bases in Watson and Crick-type pairing. However, at about 1700 nt in length, the HDV genome is much larger than that of viroid RNAs, which seem to range from about 250 to 400 nt. It contains a ribozyme domain of about 85 nt, as will be discussed in Sect. 2.2.3.

Inside infected cells there are two other HDV-related RNAs in addition to about 300,000 copies of the genome. There is about ten times less of its complement, called the antigenome (CHEN et al. 1986). This RNA is also circular in conformation and contains its own ribozyme.

On the antigenome is an open reading frame (ORF) for δAg. However, for two major reasons this antigenomic RNA cannot be translated: (a) it is circular in conformation, and (b), like the genome, it is located in the nucleus. For the

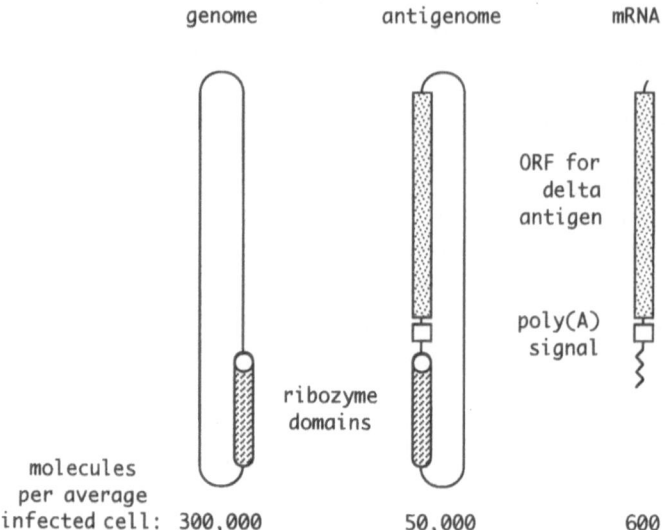

Fig. 1. Features of the three RNAs of HDV. The quantitation is from published studies (CHEN et al. 1986). *Open circles* indicate the sites of self-cleavage, which are located at the 5' ends of the ribozyme domains (PERROTTA and BEEN 1990)

translation of δAg there is yet a third HDV RNA. It is about 50 times less abundant than the antigenome, but it is cytoplasmic and 3'-polyadenylated (HSIEH et al. 1990). We do not yet know whether this RNA has a 5'-cap, but it does appropriately include the δAg ORF. It has a 5'-end near to what we refer to as the top of the rodlike structure; this topic will be discussed in Sects. 2.1.5 and 2.2.1.

In addition to these three main RNAs there are other less abundant species, such as dimers and trimers of the genome and antigenome which are by-products of replication (CHEN et al. 1986). It must also be noted that the species we readily detect are ones with an appropriate combination of (a) abundance, (b) relative stability, and (c) discrete size. Other RNAs not possessing such properties might be missed.

1.2.4 Two Delta Antigens

Before ending this introduction, it is necessary to give more consideration to δAg, the only protein encoded by HDV. Actually, there are two related forms of δAg because, during genome replication, post-transcriptional RNA editing occurs at what corresponds to the middle nucleotide of the amber termination codon. As discussed in Sect. 2.2.5, this editing ultimately allows the translation of a δAg species that is 19 amino acids (AA) longer at the C-terminus.

The original small form of δAg is 195 AA in length, and it is absolutely needed for genome replication (KUO et al. 1989). The larger form, 214 AA in length, is a dominant-negative inhibitor of genome replication (CHAO et al. 1990). Also, an unknown fraction of it becomes farnesylated at a cysteine, 4 AA from the new C-terminus (GLENN et al. 1992; OTTO and CASEY 1996). This modification is considered to be necessary for the large δAg to facilitate the assembly of new virions in cooperation with the surface proteins of the helper HBV (CHANG et al. 1991).

Both δAg proteins are highly basic, and it seems that some of the large δAg can be monophosphorylated (BICHKO et al. 1997). The small and large δAg also share three activities that have been mapped to domains (Fig. 2): (a) At position 12–60 there is a domain which mediates multimerization (ROZZELLE et al. 1995). The crystal structure indicates an antiparallel coiled-coil structure (ZUCCOLA et al. 1998). (b) Between positions 69 and 88 is a bipartite nuclear localization signal (XIA et al. 1992). (c) Next, between positions 97 and 146 is a bipartite RNA-binding domain which is sufficient for specific δAg binding to the rodlike RNAs of HDV (LEE et al. 1993). However, there other basic regions on δAg, and these may be involved in less specific interactions with nucleic acids. Recent in vitro studies suggest that the positively charged δAg can act as an efficient chaperone for nucleic acids, which are negatively charged (MORALEDA, NETTER, and TAYLOR, unpublished). The region near the C-terminus of δAg is rich in proline and glycine, but no specific function has yet been shown for this domain.

It will be important to determine the full crystal structures of the two δAgs. In addition, we will need to understand not only the protein-protein interactions involved in multimerization but also the protein-RNA interactions involved in assembly and other aspects of genome replication.

1.2.5 Concept of Primary and Secondary Accumulations

A simple scheme that has to be at the essence of what we call 'genome replication' can be summarized as follows:

– Primary accumulation = transcription + processing + stabilization
– Secondary accumulation = transcription + processing + stabilization
– Genome replication = primary accumulation + secondary accumulation

Consider first the concept of a primary accumulation. When HDV genomic RNA enters a cell and begins replication, it seems reasonable to suggest that there first must be some 'primary accumulation' of new antigenomes. Such accumulation depends upon (a) RNA-directed RNA transcription of genomic into antigenomic RNA, (b) processing of these newly synthesized RNAs so as to make new circular antigenomes, and (c) further steps of stabilization of these circles, so that they can accumulate rather than be degraded. 'Secondary accumulation' then involves the

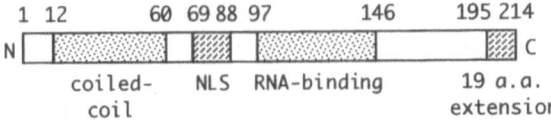

Fig. 2. Activities that can be mapped to domains on the delta antigen(s). The numbering for the domains is based on published studies (MACNAUGHTON and GOWANS 1995; ROZZELLE et al. 1995)

use of these newly accumulated antigenomes as templates for the corresponding three steps leading to accumulation of new genomes. We would say that replication has occurred if, as a result of these two accumulations, there is a net increase, with time, in the amount of genomic RNA within a cell.

This scheme is, of course, oversimplified. For example, even during the primary accumulation, a fraction of the HDV transcripts must be processed to make mRNA. This division both is essential and has to be regulated. Otherwise, there would be no opportunity to achieve even the primary accumulation of circular antigenomes.

In Sect. 2, therefore, the replication of the HDV genome, which is the topic of this article, will be considered under three headings: transcription, processing, and stabilization. As will be explained, even these headings have their limitations. For example, RNA processing per se can also be a form of RNA stabilization.

2 Genome Replication

2.1 Transcription

2.1.1 Redirection of Pol II?

As with the viroids, we have the problem of determining which enzyme(s) is involved in the RNA-directed RNA synthesis. For one viroid there is evidence that replication occurs in the nucleus using RNA polymerase II (SCHINDLER and MU-HLBACH 1992).

There are two kinds of studies which suggest that HDV, which replicates in the nucleus, could redirect the host RNA polymerase II. The first kind of study employed nuclear extracts. Studies first made by MACNAUGHTON et al. (1991) showed that extracts from cells undergoing HDV RNA synthesis continued to make HDV RNA in a reaction that was sensitive to α-amanitin, consistent with the enzyme being pol II. Later studies by FU and TAYLOR (1993), using extracts from uninfected cells, demonstrated a reaction with exogenously supplied HDV RNA that was also α-amanitin sensitive. Puzzlingly, it has not been possible to reproduce these results.

The second type of study was with purified pol II and basal transcription factors. In this study there was transcription of HDV RNA; it depended upon pol II and two basal transcription factors, TBP and TFIIB (FU, TAYLOR, SHENK, and USHEVA, unpublished).

In all of these in vitro transcription studies δAg was not needed, but it was still not excluded that in the presence of δAg there might be some qualitative or quantitative difference in the reactions. Thus, further studies are needed to test this more carefully, especially since it is quite clear that in vivo the δAg is needed for overall genome replication (KUO et al. 1989), although its role(s), as discussed in Sect. 2.1.2, might be indirectly rather than directly related to RNA transcription.

2.1.2 Role of δAg

In the overall scheme of HDV genome replication there are already data, obtained from in vivo experiments, pointing to three roles for δAg-S. (a) The presence of δAg-S can help suppress the polyadenylation of antigenomic HDV RNAs that contain rodlike structures surrounding the polyadenylation signals (see Sect. 2.2.2). (b) When circular forms of HDV RNA are produced within a cell, the presence of δAg can help stabilize these RNAs (LAZINSKI and TAYLOR 1994). The amount of accumulation can be increased by the presence of either the small form, δAg-S, or the large form, δAg-L (see Sect. 2.3.2). (c) It has been shown that even though the HDV ribozymes do function within cells in the absence of δAg, there is an enhanced level of cleavage when δAg is present (see Sect. 2.2.3). A fourth role of δAg-S might be in the transport of HDV RNA to the nucleus to facilitate transcription. Consistent with this speculation are the results that δAg-S has both RNA-binding activity and also a nuclear localization signal. Recently, more direct supporting evidence has come from studies in which permeabilized cells will take up HDV RNA into the nuclei if δAg-S is present (CHOU et al. 1998).

2.1.3 Viroid-like Domain?

For a long time it has been tempting, based upon the analogy to the plant viroids, to consider the HDV genome as being composed of two parts: first, a smaller viroid-like domain and second, a larger part containing the ORF for δAg. At this point the model of a viroid-like domain for HDV has several problems, one of which is that no one has been able to reduce the HDV RNA to a viroid-sized minigenome and achieve replication, independent of whether or not δAg-S is provided separately (LAZINSKI and TAYLOR 1994). There is a report from Brazas and Ganem that normal cells contain a protein, called DIPA, that can bind δAg and might also be a cellular homologue of δAg (BRAZAS and GANEM 1996). The claim of homology remains controversial (BRAZAS and GANEM 1997; LONG et al. 1997), but it has stimulated speculation that HDV arose via some kind of recombination between a putative ancestral human viroid and the mRNA for DIPA (ROBERTSON 1996).

2.1.4 Promoters?

The next question about HDV transcription is whether or not there are promoters for RNA-directed RNA synthesis. One hypothesis is that the 5′-end of the mRNA should be in the vicinity of such a promoter element. Furthermore, two studies have led to the additional hypothesis that such an element looks like a DNA-directed pol II promoter. In their initial studies, investigators asked whether a double-stranded cDNA copy of sequences near the top of the rod would be active when placed in front of a reporter gene in a promoterless vector. In this way, a region was found which would act as a modest promoter in either orientation; when the element was reduced to a little as 29 nt it was still functional, although now only in one orientation (MACNAUGHTON et al. 1993). One could question the logic of using the double-stranded cDNA to understand what happens for a template that is not DNA but RNA, and not 100% but only partially double stranded.

Nevertheless, evidence has been obtained by two other approaches to support this hypothesis of a promoter in this region. The first is phylogenetic. Beard et al. reported a comparison of several different HDV genomes in this region (BEARD et al. 1996). They identified several short conserved sequences in this region, and of these they emphasized a short paired region rich in guanosine and cytosine. The second approach was to make a series of mutations in the RNA genome and measure the consequences for the accumulation of stable processed RNA (BEARD et al. 1996; WANG et al. 1997; WU et al. 1997). However, in the cases where inhibition was observed, it soon became apparent that such an assay, dependent on the multiple steps of primary and secondary accumulation, as defined earlier (Sect. 1.2.5 and Table 1), was too nonspecific to be an indicator of what was needed for the initiation of RNA-directed RNA synthesis (WU et al. 1997). For example, in some cases a 50-fold inhibition of accumulation was observed early, but with time, a kind of 'catch-up' occurred in which the mutant genomes actually accumulated to final levels close to those achieved in less time by the unmutated wild-type genome. As another example, some mutants with an apparent 100-fold inhibition and no indication of catch-up, were tested in cells already expressing the small δAg, and it was found that the replication was now more extensive, reaching levels close to wild type.

It was thus clear that RNA accumulation was more an overall rather than a specific indicator of the consequences of genome modification. Thus, as described next in Sect. 2.2.2, we focused on the 5'-ends detected for the polyadenylated HDV RNAs.

2.1.5 5'-Ends and Initiation

As mentioned in Sect. 1.2.3, the finding of mRNA species with a 5' end near the top of the rod has led to tests of the hypothesis that the 5' end corresponds to a site of initiation. We recently used a version of 5'-RACE (rapid amplification of cDNA ends) to better characterize such 5' ends not only during wild-type replication but also during the replication of mutated viral genomes (DINGLE, GUDIMA, WU, TAYLOR, unpublished).

For the replication of wild-type genomes both in an infected woodchuck liver and also in transfected cells, the 5' end corresponds to nucleotide 1630, in close agreement with earlier primer extension studies which mapped the 5' end to nt 1631 ± 1 (HSIEH and TAYLOR 1991). If 1630 were actually an initiation site then transcription would begin with adenosine and correspond to a site in the middle of a 3-nt external bulge on the predicted rodlike structure. The same 5'-RACE was also applied to many different mutated genomes. Not surprisingly, actual mutagenesis at and around nucleotide 1630 sometimes moved the 5' end away from this site. However, mutagenesis at some more distant sites (relative to 1630) which led to inhibition of genome accumulation also gave 5' sites other than 1630. It was almost as if those mutants with lower levels of RNA accumulation gave more 5' sites. Intriguingly, when wild-type infections were studied early, before genome replication was well underway, the same kind of heterogeneity in 5' ends was observed.

One hypothesis currently being tested is that the amount of available δAg can increase the apparent specificity of the 5'-end formation. It might be that δAg can form a complex on the RNA and somehow help define where initiation takes place. However, the 5' ends detected may correspond to a population of more stable RNAs; there may be initiations at other sites, but the RNAs produced are less stable. It should also be noted that there are no data regarding the number or location of sites for the initiation of genomic RNA from antigenomic RNA templates.

2.2 Processing

2.2.1 5'-Capping?

There are several RNA processing steps that are critical to HDV genome replication. 5'-Capping of the mRNA is probably one of these steps, but as of this time, there is no direct evidence for or against capping. The only evidence for 5'-capping remains circumstantial; i.e., we know that many linear mRNAs are stabilized in part by a 5'-cap structure, and so we expect that this is also true for the HDV mRNA.

2.2.2 Polyadenylation

The processing of the 3' ends of the HDV mRNA seems to occur via the standard cellular machinery usually applied to RNAs transcribed by pol II. We can recognize three standard sequence features on the antigenomic RNA: (a) a polyadenylation signal, AAUAAA, (b) a CA acceptor sequence, and (c) a downstream G/U-rich region (HSIEH et al. 1990). Since all these features seem ideal for polyadenylation, the process must be negatively regulated. Otherwise, there will not be any unprocessed full-length antigenomes.

Studies using cDNA-transfected cells suggest two factors that may contribute to the negative regulation (HSIEH and TAYLOR 1991). (a) If a nascent antigenomic RNA is able to fold into the rodlike structure prior to polyadenylation, then the polyadenylation may be inhibited. Such folding would not be possible for the 800 base mRNA. However, in a rolling-circle model of replication, such as shown in Fig. 3, if there were transcripts that proceeded beyond the polyadenylation region, then by the time they reached that region for a second time, they would be able to so fold. (b) As mentioned earlier, the binding of δAg to such a rodlike structure might provide further inhibition of polyadenylation. These two forms of regulation have been shown to apply in vivo to DNA-directed HDV transcripts (HSIEH and TAYLOR 1991), but there are, as yet, no data to prove the same occurs during natural genome replication.

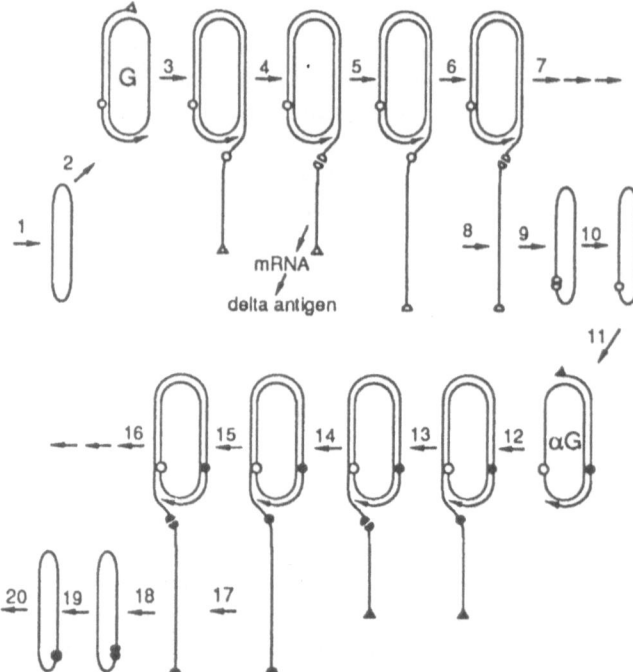

Fig. 3. A double rolling-circle model of HDV genome replication. The diagram shows how a unit-length circular genomic RNA might be transcribed into a polyadenylated mRNA and continue on to produce multimers of antigenomic RNA, which undergo self cleavage and then ligation to make new unit-length circular antigenomic RNAs (*steps 1–10*). The subsequent process of converting such antigenomes into genomes is simpler since there would not be any processing by polyadenylation (*steps 11–20*). The sites for initiation and self-cleavage are indicated by *small triangles* and *circles*, respectively. The *closed* and *open symbols* refer to genomic (*G*) and antigenomic (*αG*) species, respectively. (Reprinted with permission from Taylor et al. 1989)

2.2.3 Cleavage

The next form of RNA processing related to genome replication is RNA cleavage. Both the genome and antigenome undergo self-cleavage in vitro. These two reactions are dependent upon a divalent metal ion, such as magnesium (Kuo et al. 1988; Sharmeen et al. 1988). Also, the cleavage produces via a transesterification reaction a 5'-OH and a 2'-, 3'-cyclic monophosphate. In this respect the HDV ribozymes are like those of the plant agents. However, when it comes to the relevant RNA folding, the delta ribozymes are different from anything described so far (Symons 1997).

Several ribozyme features need to be noted. (a) The minimum amount of contiguous sequence seems to be only 85 nt (Perrotta and Been 1990). (b) The folding of this RNA involves more than just regular Watson and Crick pairing. There is evidence for a pseudoknot and also for unusual pairings, such as a G-G pair (Wickham et al. 1997). (c) Recent studies have described the successful crystallization of an HDV ribozyme domain (Ferré-D'Amaré et al. 1998) and there is an as yet unpublished claim for a detailed structure. (d) While the HDV cleavage reaction can occur in vitro in the absence of added protein, the story in vivo might be different. There is evidence that the presence of δAg can enhance the accumulation of the processed RNA (JENG et al. 1996). It is as if δAg can act as an RNA chaperone.

2.2.4 Ligation

The next processing reaction is ligation to form circles. As mentioned above, the cleavage reactions of HDV RNAs, just like those for the plant viroids, produce ends with a 5'-OH and a 2'-, 3'-cyclic monophosphate. It has been known for many years that when such ends are brought together via a complementary guide sequence, ligation will occur in the absence of any added protein (USHER and McHALE 1976). Such 'self-ligation', with an RNA guide sequence, was also shown for HDV RNAs (SHARMEEN et al. 1989).

Some years ago, the concept of guide-directed self-ligation was incorporated into a double rolling-circle model of HDV genome replication, as shown in Fig. 3. It was speculated that the other side of the rodlike structure would act, via its 70% base pairing, as a guide to facilitate self-ligation (TAYLOR et al. 1987). Even though this aspect of the model seems appealing, there are now some data to the contrary. Experiments using HDV RNAs expressed in transfected cells show that the rodlike structure is not needed for ligation (LAZINSKI and TAYLOR 1995a). This result raises the possibility that maybe inside cells, cleaved HDV RNAs are not self-ligated but depend upon a host ligase (LAZINSKI and TAYLOR 1995b). Consistent with this, Neel and Robertson report that HeLa nuclear extracts, which contain two different host tRNA ligases, can ligate HDV RNAs (NEEL and ROBERTSON 1997).

2.2.5 Editing

As mentioned in Sect. 1.2.4, one of the consequences of this RNA-editing is for the termination codon of δAg-S to be replaced by tryptophan, allowing for an extra 19 AA to be added to create the δAg-L (LUO et al. 1990). The mechanism of this editing is becoming clearer: (a) The substrate for the change is the adenosine at nt 1012 on the antigenome (CASEY and GERIN 1995; POLSON et al. 1996). (b) An adenosine deaminase converts this adenosine to inosine. Following additional rounds of genome replication, the adenosine is replaced by guanosine. (c) The actual enzyme has not yet been identified, but it is probably one of a family of double-stranded RNA-activated adenosine deaminases that are currently being referred to as ADAR (BASS et al. 1997). Figure 4 shows the predicted folding of the region on the antigenome around the adenosine that is edited. The nucleotide itself is not predicted to be paired. These ADAR are probably present in the nuclei of all mammalian cell types. They act on specific mRNA and even tRNAs (CATTANEO 1994), and they are known to act on viral RNAs other than HDV. Examples include retroviruses, vesicular stomatitis virus, and the two most striking examples:

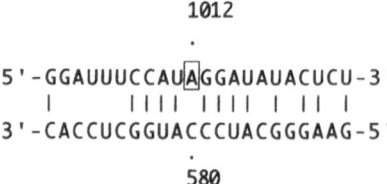

Fig. 4. Site on predicted rodlike folding of HDV antigenome at which RNA editing occurs. Nucleotide 1012 on the antigenome is deaminated to inosine (POLSON et al. 1996b)

the RNAs of polyoma virus and measles virus (CATTANEO 1994; KUMAR and CARMICHAEL 1997).

There is evidence that HDV RNAs can be edited at other sites on both the genome and the antigenome (Wu et al. 1995). One interpretation for why the editing at other sites is apparently not as frequent is that such editing, for one reason or another, is not consistent with genome replication that is as active as that of the wild-type genome. However, it needs to be pointed out that even the change at nt 1012 is, in another respect, not consistent with further genome replication, since on entering a new host cell such a genome could not encode δAg-S that is essential for genome replication. For this reason, the ADAR editing of HDV must somehow be regulated so that some but not all genomes become changed. We still do not understand how this regulation is achieved, although it may be possible that somehow the binding of δAg to the RNA inhibits editing by ADAR (CASEY and GERIN 1995).

2.3 Stabilization

2.3.1 Intrinsic Stability

The three main RNAs of HDV replication (Fig. 1) already have a level of stability based upon post-transcriptional RNA processing events that they have undergone. Specifically, the genome and antigenome are circular; there is already good evidence from in vitro and in vivo studies that circular RNAs are more stable than corresponding linear RNAs (PUTTARAJU and BEEN 1995). The reason seems to be that cellular RNases are largely exo- rather than endonucleases. Also, the linear mRNA gains stability from the presence of a 3′ tail of poly(A), and very likely, though not proven, the 5′ end is stabilized by the presence of a cap structure.

2.3.2 δAg Binding

Even though circular RNAs are more stable than corresponding linear RNAs, there is evidence that the circles of HDV require additional stabilization. It was found that when circular HDV RNAs of full size and smaller were expressed inside cells (in the absence of genome replication), the full-size circles were stabilized 15-fold by the presence of δAg (LAZINSKI and TAYLOR 1995a). As the RNAs became smaller, this effect was reduced. Interestingly, species of 358 nt, the size of a plant viroid, did not need δAg at all.

2.3.2 Attenuator

At another level, the two circular RNAs of HDV need to be stabilized against self-cleavages. Specifically, each contains all the sequences of a ribozyme and thus could cleave to produce a linear RNA. As was shown, the rodlike structure itself stops the ribozyme from folding into an active conformation (LAZINSKI and TAYLOR 1995a). That is, it is considered that the sequences on the rod opposite the ribozyme domain act as an attenuator.

3 Outlook

The perspective of this review has been to point out not only the things we do not know but also that troubles can arise from the things that we think we know. There are still likely to be some major surprises as we try to explain the HDV genome replication. Probably the mechanism of RNA-directed RNA synthesis remains the most interesting question.

At this time the structure and replication of the HDV genome remain unique among the viruses of animals. As discussed in Sect. 2.1.3, there are similarities between HDV and the plant viroids, and it has been profitable to test the extent of this analogy and make extrapolations. However, it is now becoming clearer that there is a significant diversity among the small single-stranded circular RNAs of viroids and viroid-like agents. Consider three examples. (a) While some viroids, like potato spindle tuber viroid, are considered to be replicated in the nucleus by the host RNA polymerase II (SCHINDLER and MUHLBACH 1992), there are others, like avocado sunblotch viroid, which seem to be replicated not by pol II and are de-tected in chloroplasts (LIMA et al. 1994). (b) While for most viroids replication involves RNA-directed RNA synthesis, there is one viroid-like element which is replicated via the reverse transcriptase of a helper virus to produce a double-stranded DNA intermediate; the helper virus is carnation etch ring virus which, intriguingly, has many similarities to HBV, the helper virus of HDV (DAROS and FLORES 1995). (c) Finally, in the mitochondria of *Neurospora* there can be the accumulation of a not so small circular RNA (881 nt) that is replicated via reverse transcription directed by a helper plasmid (KENNELL et al. 1995). For the future, we might have to take a wider view in our attempts to extrapolate and find the roots of HDV.

Acknowledgements. In order to assemble this overview I have presented published and unpublished work from both my own and other labs. Apologies are offered in advance for cases where acknowledgements are omitted and, more importantly, where misinterpretations are given. I would like to thank the *Instituto Juan March de Estudios e Investigaciones* for sponsoring a workshop entitled "Plant Viroids and Viroid-like Satellite RNAs from Plants, Animals and Fungi", held December 1 3, 1997, in Madrid, that was very helpful for the preparation of this manuscript. Thanks are also due to William Mason, Ting-Ting Wu, and Kate Dingle for their constructive comments on the manuscript.

References

Bass BL, Nishikura K, Keller W, Seeburg PH, Emeson RB, O'Connell MA, Samuel CE, Herbert A (1997) A standardized nomenclature for adenosine deaminases that act on RNA. RNA 3:947 949

Beard MR, Macnaughton TB, Gowans EJ (1996) Identification and characterization of a hepatitis delta virus RNA transcriptional promoter. J Virol 70:4986 4995

Bichko V, Barik S, Taylor J (1997) Phosphorylation of the hepatitis delta virus antigens. J Virol 71:512 518

Brazas R, Ganem D (1996) A cellular homolog of hepatitis delta antigen: implications for viral replication and evolution. Science 274:90 94

Brazas R, Ganem D (1997) Delta-interacting protein A and the origin of hepatitis delta antigen. Science 276:825

Casey JL, Gerin JL (1995) Hepatitis D virus RNA editing: specific modification of adenosine in the antigenomic RNA. J Virol 69:7593 7700

Cattaneo R (1994) Biased (adenosine to inosine) hypermutation in animal virus genomes. Curr Opin Genet Dev 4:895 900

Chang FL, Chen PJ, Tu SJ, Chiu MN, Wang CJ, Chen DS (1991) The large form of hepatitis δ antigen is crucial for the assembly of hepatitis δ virus. Proc Natl Acad Sci USA 88:8490 8494

Chao M, Hsieh S-Y, Taylor J (1990) Role of two forms of the hepatitis delta virus antigen: evidence for a mechanism of self-limiting genome replication. J Virol 64:5066 5069

Chen P-J, Kalpana G, Goldberg J, Mason W, Werner B, Gerin J, Taylor J (1986) Structure and replication of the genome of hepatitis δ virus. Proc Natl Acad Sci USA 83:8774 8778

Chou H-C, Hsieh T-Y, Sheu G-T, Lai MMC (1998) Hepatitis delta antigen mediates the nuclear import of hepatitis delta virus RNA. J Virol 72:3684 3690

Daros J, Flores R (1995) Identification of a retroviroid-like element from plants. Proc Natl Acad Sci USA 92:6856 6860

Diener TO, Prusiner SB (1985) The recognition of subviral pathogens. In: Marmorosch K, McKelvey JJ jr (eds) Subviral pathogens of plants and animals: viroids and prions. Academic, London

Ferré-D'Amaré AR, Zhou K, Doudna JA (1998) A general module for RNA crystallization. J Mol Biol 279:621 631

Fu T-B, Taylor J (1993) The RNAs of hepatitis delta virus are copied by RNA polymerase II in nuclear homogenates. J Virol 67:6965 6972

Glenn JS, Taylor JM, White JM (1990) In vitro-synthesized hepatitis delta virus RNA initiates genome replication in cultured cells. J Virol 64:3104 3107

Glenn JS, Watson JA, Havel CM, White JO (1992) Identification of a prenylation site in the delta virus large antigen. Science 256:1331 1333

Hadziyannis SJ (1997) Review: hepatitis delta virus. J Gastroenterol Hepatol 12:289 298

Hsieh S-Y, Taylor J (1991) Regulation of polyadenylation of HDV antigenomic RNA. J Virol 65:6438 6446

Hsieh S-Y, Chao M, Coates L, Taylor J (1990) Hepatitis delta virus genome replication: a polyadenylated mRNA for delta antigen. J Virol 64:3192 3198

Jeng K-S, Su P-Y, Lai MMC (1996) Hepatitis delta antigen enhances the ribozyme activities of hepatitis delta virus RNA in vivo. J Virol 70:4205 4209

Kennell JC, Saville BJ, Mohr S, Kuiper MTR, Sabourin JR, Collins RA, Lambowitz AM (1995) The VS catalytic RNA replicates by reverse transcription as a satellite of a retroplasmid. Genes Dev 9:294 303

Kumar M, Carmichael GG (1997) Nuclear antisense RNA induces extensive adenosine modifications and nuclear retention of target transcripts. Proc Natl Acad Sci USA 94:3542 3547

Kuo MYP, Sharmeen L, Dinter-Gottlieb G, Taylor J (1988) Characterization of self-cleaving RNA sequences on the genome and antigenome of human hepatitis delta virus. J Virol 62:4439 4444

Kuo MYP, Chao M, Taylor J (1989) Initiation of replication of the human hepatitis delta virus genome from cloned DNA: role of delta antigen. J Virol 63:1945–1950

Lai MMC (1995) The molecular biology of hepatitis delta virus. Annu Rev Biochem 64:259–286

Lazinski DW, Taylor JM (1994) Expression of hepatitis delta virus RNA deletions: cis and trans requirements for self-cleavage, ligation, and RNA packaging. J Virol 68:2879–2888

Lazinski DW, Taylor JM (1995a) Intracellular cleavage and ligation of hepatitis delta virus genomic RNA: regulation of ribozyme activity by cis-acting sequences and host factors. J Virol 69:1190–1200

Lazinski DW, Taylor JM (1995b) Regulation of the hepatitis delta virus ribozymes: to cleave or not to cleave. RNA 1:225–233

Lee C-Z, Lin J-H, McKnight K, Lai MMC (1993) RNA-binding activity of hepatitis delta antigen involves two arginine-rich motifs and is required for hepatitis delta virus RNA replication. J Virol 67:2221 2227

Lima MI, Fonseca ME, Flores R, Kitajima EW (1994) Detection of avocado sunblotch viroid in chloroplasts of avocado leaves by in situ hybridization. Arch Virol 138:385 390

Long M, de Souza SJ, Gilbert W (1997) Delta-interacting protein A and the origin of hepatitis delta antigen. Science 276:824 825

Luo G, Chao M, Hsieh S-Y, Sureau C, Nishikura K, Taylor J (1990) A specific base transition occurs on replicating hepatitis delta virus RNA. J Virol 64:1021–1027

Macnaughton T, Gowans E (1995) Hepatitis delta antigen. In: Dinter-Gottlieb G (ed) The unique hepatitis delta virus. Landes, Austin

Macnaughton TB, Gowans EJ, McNamara SP, Burrell CJ (1991) Hepatitis delta antigen is necessary for access of hepatitis delta virus RNA to the cell transcriptional machinery but is not part of the transcriptional complex. Virology 184:387 390

Macnaughton TB, Beard MR, Chao M, Gowans EJ, Lai MMC (1993) Endogenous promoters can direct the transcription of hepatitis delta virus RNA from a recircularized cDNA template. Virology 196:629 636

Margolis HS, Alter MJ, Hadler SC (1997) Viral hepatitis. In: Evans AS, Kaslow RA (eds) Viral infections of humans: epidemiology and control. Plenum, New York

Neel OD, Robertson HD (1997) Pathways of RNA ligation in the replication of the hepatitis delta agent. Nucleic Acids Res Symp Ser 36:154–155

Otto JC, Casey PJ (1996) The hepatitis delta virus large antigen is farnesylated both in vitro and in animal cells. J Biol Chem 271:4569–4572

Perrotta AT, Been MD (1990) The self-cleaving domain from the genomic RNA of hepatitis delta virus: sequence requirements and the effects of denaturant. Nucleic Acids Res 18:6821–6827

Polson AG, Bass BL, Casey JL (1996) RNA editing of hepatitis delta virus antigenome by dsRNA-adenosine deaminase. Nature 380:454–456

Ponzetto A, Cote PJ, Popper H, Hoyer BH, London WT, Ford EC, Bonino F, Purcell RH, Gerin JL (1984) Transmission of the hepatitis B virus-associated δ agent to the eastern woodchuck. Proc Natl Acad Sci U S A 81:2208 2212

Ponzetto A, Negro F, Purcell RH (1991) Experimental hepatitis delta virus infection in the animal model. In: Gerin JL, Purcell RH, Rizzetto M (eds) The hepatitis delta virus. Liss, New York

Purcell RH, Gerin JL (1996) Hepatitis delta virus. In: Fields BN, Knipe DM, Howley PM (eds) Fields virology. Raven, New York

Puttaraju M, Been M (1995) Generation of nuclease-resistant circular RNA decoys for HIV-tat and HIV-rev by autocatalytic splicing. Nucleic Acids Res 33:49–51

Rizzetto M, Canese MG, Arico J, Crivelli O, Bonino F, Trepo CG, Verme G (1977) Immunofluorescence detection of a new antigen-antibody system associated to the hepatitis B virus in the liver and in the serum of HBsAg carriers. Gut 18:997–1003

Rizzetto M, Canese MG, Gerin JL, London WT, Sly DL, Purcell RH (1980) Transmission of the hepatitis B virus-associated delta antigen to chimpanzees. J Infect Dis 141:590–602

Robertson HD (1996) How did replicating and coding RNAs first get together? Science 274:66 67

Rozzelle J, Wang J-G, Wagner D, Erickson B, Lemon S (1995) Self-association of a synthetic peptide from the N terminus of the hepatitis delta virus protein into an immunoreactive alpha-helical multimer. Proc Natl Acad Sci 92:382–386

Schindler I-M, Muhlbach H-P (1992) Involvement of nuclear DNA-dependent RNA polymerases in potato spindle tuber viroid replication: a reevaluation. Plant Sci 84:221 229

Sharmeen L, Kuo MY, Dinter-Gottlieb G, Taylor J (1988) The antigenomic RNA of human hepatitis delta virus can undergo self-cleavage. J Virol 62:2674–2679

Sharmeen L, Kuo MY, Taylor J (1989) Self-ligating RNA sequences on the antigenome of human hepatitis delta virus. J Virol 63:1428–1430

Symons RH (1997) Plant pathogenic RNAs and RNA catalysis. Nucleic Acids Res 25:2683–2689

Taylor J, Kuo M, Chen P-J, Kalpana G, Goldberg J, Aldrich C, Coates L, Mason W, Summers J, Gerin J, Baroudy B, Gowans E (1987) Replication of hepatitis delta virus. In: Robinson W, Koike K, Will H (eds) Hepadna viruses. Liss, New York

Taylor J, Sharmeen L, Kuo M, Dinter-Gottlieb G (1989) The self-cleaving RNAs of human hepatitis delta virus. In: Cech T (ed) Molecular biology of RNA. UCLA symposia on molecular and cellular biology. Liss, New York

Taylor J, Chao M, Kuo M, Sharmeen L, Hsieh S-Y (1990) Human hepatitis delta: unique or not unique. In: Brinton M, Heinz FX (eds) New aspects of positive-strand viruses. ASM Publications, Washington

Usher D, McHale A (1976) Nonenzymic joining of oligoadenylates on a polyuridylic acid template. Science 192:53–54

Wang H-W, Wu H-L, Chen D-S, Chen P-J (1997) Identification of the functional regions required for hepatitis D virus replication and transcription by linker-scanning mutagenesis of viral genome. Virology 239:119–131

Wang K-S, Choo Q-L, Weiner AJ, Ou J-H, Najarian C, Thayer RM, Mullenbach GT, Denniston KJ, Gerin JL, Houghton M (1986) Structure, sequence and expression of the hepatitis delta viral genome. Nature 323:508–513

Wickham GS, Shih I-H, Been MD (1997) Molecular modeling of a G-G base pair in in the antigenomic HDV ribozyme. Nucleic Acids Res Symp Ser 36:99–101

Wu T-T, Bichko VV, Ryu W-S, Lemon SM, Taylor JM (1995) Hepatitis delta virus mutant: effect on RNA editing. J Virol 69:7226–7231

Wu T-T, Netter HJ, Lazinski DW, Taylor JM (1997) Effects of nucleotide changes on the ability of hepatitis delta virus to transcribe, process and accumulate unit-length, circular RNA. J Virol 71:5408–5414

Xia Y-P, Yeh C-T, Ou J-S, Lai MMC (1992) Characterization of nuclear targeting signal of hepatitis delta antigen: nuclear transport as a protein complex. J Virol 66:914–921

Zuccola HJ, Rozzelle JE, Lemon SM, Erickson BW, Hogle JM (1998) Structural basis of the oligomerization of hepatitis delta antigen. Structure 6:821–830

Biology and Structure of Plant Satellite Viruses Activated by Icosahedral Helper Viruses

K.-B.G. Scholthof,[1] R.W. Jones,[2] and A.O. Jackson[3]

[1]Department of Plant Pathology and Microbiology, Texas A&M University, College Station, TX 77843, USA
[2]USDA-ARS, BARC-West, Bldg. OIOA, 10300 Baltimore Ave., Beltsville, MD 20705, USA
[3]Department of Plant and Microbial Biology, University of California, Berkeley, CA 94720, USA
Research support was obtained through funding from the Texas Agricultural Experiment Station Grant H-8388 and USDA Competitive Grant 96 35303 3714 to K.-B.G.S. and USDA Grant 87-CRCR-2556 to A.O.J.

1 Introduction

Nearly 40 years ago, KASSANIS and NIXON (1960) described two kinds of particles in tobacco necrosis virus (TNV) infections that differed in size, and thus seemed to be separate viruses. The smaller component was demonstrated to depend on the larger autonomous virus for its multiplication, and hence was designated a satellite virus, satellite tobacco necrosis virus (STNV). This discovery quickly stimulated research in several areas of biology and chemistry with a focus on relationships, pathology, transmission, and interdependence of STNV and TNV. Following the realization of the satellite nature of STNV, identification of other satellite viruses was remarkably slow compared with the rates of discovery of new viruses, satellite RNAs, and viroids. More than 15 years passed before a second satellite virus, satellite panicum mosaic virus (SPMV; Fig. 1), was described (BUZEN et al. 1984; NIBLETT and PAULSEN 1975) . A satellite virus particle associated with maize white line mosaic virus (MWLMV) that constitutes a third distinct satellite virus was subsequently identified (GINGERY and LOUIE 1985; ZHANG et al. 1991a) . The helper viruses that support replication of these three satellite viruses are all isometric and have monopartite genomes. However, particle morphology of the helper virus seems to be inconsequential for support of isometric satellite viruses, since a fourth satellite virus, satellite tobacco mosaic virus (STMV), naturally occurs in *Nicotiana glauca* infected with the rod-shaped tobacco mosaic virus (VALVERDE and DODDS 1986). The fact that only four satellite viruses supported by RNA helper viruses have been identified to date, despite extensive research on virus characterization over the past 40 years, suggests that these agents are not widespread among the plant viruses. In this chapter, we will discuss the biology of satellite viruses supported by small polyhedral helper viruses and, where appropriate, we will compare their structure and physicochemical properties. The properties of STMV are discussed in the accompanying chapter by Dodds.

Plant satellite viruses encapsidate a positive-sense single-stranded RNA (ssRNA) genome that is distinct from and much smaller than the respective helper virus RNA genome. The coat proteins (CPs) encoded by satellite viruses are serologically unrelated to the helper virus CPs, and their virions also differ in both their architecture (Fig. 1) and in the complexity of their encapsidated RNAs (Figs. 2, 3). Another important feature is that all described helper viruses replicate independently of their associated satellite agents, and some of these also support replication of another distinct entity, a satellite RNA (sat-RNA). However, sat-RNAs are readily distinguished from satellite viruses because they do not encode a CP, and they normally are encapsidated by the helper virus CPs (FRANCKI 1985; ROOSSINCK et al. 1992). The effects of satellite virus-helper interactions on the host plant are varied, and these interactions may either exacerbate or attenuate symptoms induced by the helper virus. These properties are reviewed below.

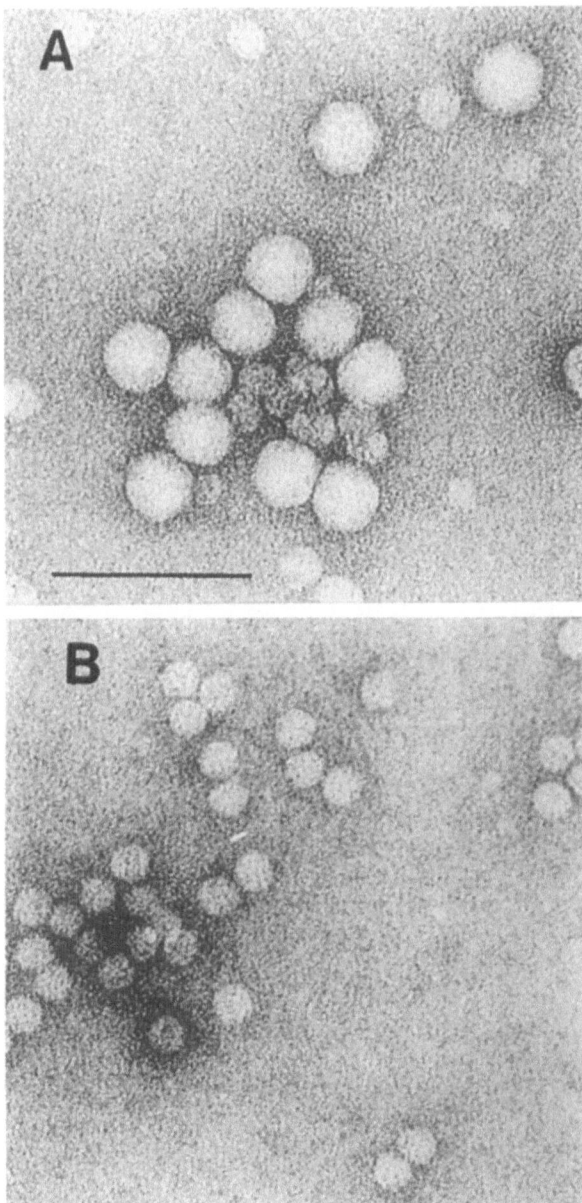

Fig. 1A,B. Electron micrographs of panicum mosaic virus (PMV) and its satellite virus (SPMV). **A** A mixture of 30 nm PMV and 17 nm SPMV particles that were present in preparations obtained from a mixed infection of pearl millet. **B** The 42 S SPMV particles following separation from the 109 S PMV particles on sucrose density gradients. The *bar* represents 100 nm

2 Biology and Pathology

2.1 Satellite Tobacco Necrosis Virus

STNV was the first satellite virus to be described in the early 1960s, and its unique properties catalyzed a barrage of investigations with a wide range of plant-infectious agents that ultimately resulted in the identification of multicomponent viruses, satellite RNAs, and viroids. A large number of studies on STNV described in various reports have also revealed aspects of helper-satellite relationships, fungal transmission, genetics, and interdependence of STNV and TNV on the disease phenotype (Danthinne et al. 1991; Kassanis 1962; Meulewaeter et al. 1993; Rees et al. 1970; Uyemoto et al. 1968; van Emmelo et al. 1987). The helper virus, TNV, can be transmitted experimentally to a wide range of host plants by me-

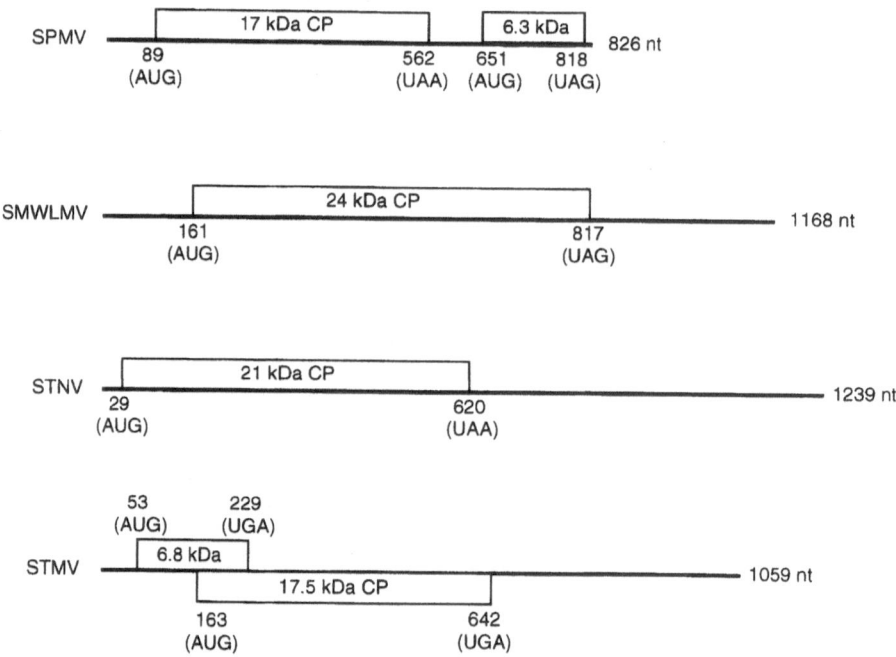

Fig. 2. Schematic organization of the plant satellite virus genomes. SPMV RNA encodes the 17-kDa capsid protein (CP) at the 5′-proximal region of the genome and a 3′-proximal 6.3-kDa open reading frame (ORF) of unknown function (Matsuta et al. 1987). Two small ORFs are also located on the minus strand (not shown). The satellite virus of MWLMV (SMWLMV) encodes a 24-kDa CP (Zhang et al. 1991b) and the TNV satellite viruses, STNV-1 and STNV-2, encode capsid proteins of 21 kDa (van Emmelo et al. 1987; Danthinne et al. 1991). The 3′ noncoding regions of the satellite RNAs are highly structured and this extensive base pairing probably reflects *cis*-requirements for replication, translation of the encoded genes, and packaging by the capsid protein. STMV, the satellite virus of TMV, is also included for comparison (Mirkov et al. 1989). The genome lengths of the satellite virus RNAs are indicated in nucleotides (*nt*). Further details on STMV are presented in the accompanying chapter by Dodds

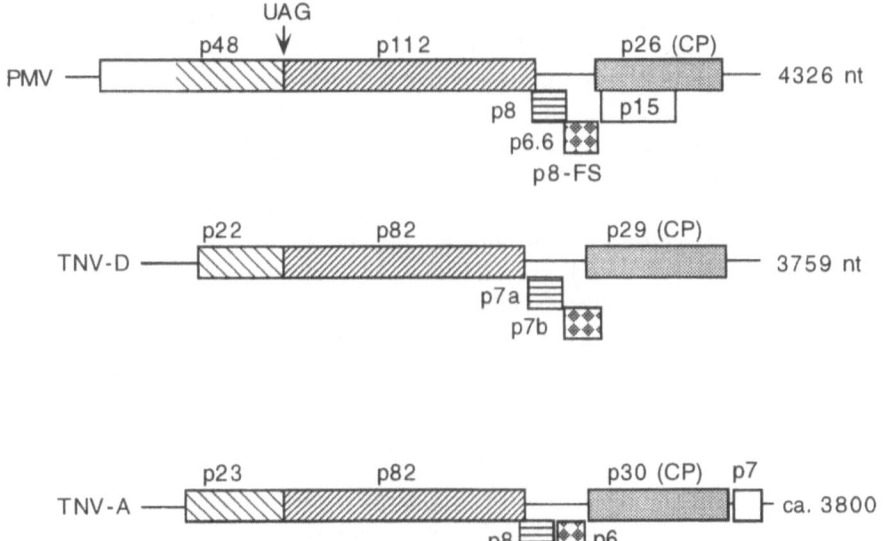

Fig. 3. Genome organization of the single-stranded RNA helper viruses associated with SPMV and STNV. Panicum mosaic virus (*PMV*; TURINA et al. 1998) and tobacco necrosis virus strain A (*TNV-A*; MEULEWAETER et al. 1990) and strain D (*TNV-D*; COUTTS et al. 1991; MOLNAR et al. 1997) are members of the *Tombusviridae*. The genomes are aligned by the putative replicase readthrough codon (*UAG*). Boxes with similar patterns indicate open reading frames (ORFs) with amino acid sequence relatedness. The *white boxes* depict PMV p15 and TNV-A p7, which are not related, and the coat proteins (*CP*) of the helper viruses are represented by the *shading patterns*. The approximate sizes of the encoded proteins are indicated in kilodaltons and the lengths of the genomes are indicated. The 3'-terminal RNA sequence of TNV-A has not been determined; the genome length is approximately 3800–4000 nt (MEULEWAETER et al. 1990)

chanical inoculation. However, soil-borne transmission of TNV by the chytrid fungus, *Olpidium brassicae*, is the principle means of natural spread of the virus (TEAKLE 1962) . STNV is also vectored efficiently by *Olpidium*, and most serotypes can be mechanically transmitted in association with the helper virus (REES et al. 1970).

STNV was initially thought to be a degradation product of the larger TNV particle but was later found to be serologically distinct from and dependent on the larger particle for its replication (KASSANIS 1962; KASSANIS and NIXON 1960). The fact that the smaller virus failed to multiply unless inoculated in the presence of the larger particle, led to the satellite–helper virus concept (KASSANIS 1962, 1981). The biological consequence of STNV is its ability to effectively "parasitize" the helper virus by competing for utilization of components required for viral replication, and this feature normally results in a pronounced attenuation of the symptoms induced by TNV. In fact, symptoms can be reduced to such an extent that infected plants growing in soils containing both TNV and STNV can escape detection during visual surveys (KASSANIS and PHILLIPS 1970).

Several serotypes of STNV and TNV have been reported. Two major groups or serotypes of TNV, designated A and D, were described initially (BABOS and

KASSANIS 1963b), and the nucleotide sequences of both strains have since been determined (COUTTS et al. 1991; MEULEWAETER et al. 1990; MOLNAR et al. 1997). Three major strains of STNV from Europe and North America have been defined based on serological analyses, amino acid composition of the capsid proteins, host range, fungal vector specificity, and the combination of TNV isolates that support their replication (KASSANIS and MACFARLANE 1968; KASSANIS and PHILLIPS 1970; REES et al. 1970; UYEMOTO et al. 1968). STNV-1 and STNV-2 are supported by TNV strains that replicate readily in tobacco and French bean following mechanical inoculation. The third strain, STNV-C, does not multiply in the inoculated leaves of these plants and must be maintained on roots of tobacco plants infected with *O. brassicae* (REES et al. 1970). These findings suggest that the host plant and helper virus have considerable effects on the replication of each particular satellite virus. They also raise questions about whether the satellite genomes and/or capsid proteins may have evolved to interact with host-specific factors to promote interactions with the helper virus. Remarkably, these questions were first raised by KASSANIS during the early stages of STNV characterization and its effects on TNV disease development (KASSANIS 1962; KASSANIS and MACFARLANE 1965; KASSANIS and NIXON 1960; REES et al. 1970).

Evidence for fungal transmission of TNV and STNV by the chytrid fungus *Olpidium brassicae* has been obtained in different laboratories (CAMPBELL and FRY 1966; KASSANIS and MACFARLANE 1968; TEAKLE 1962; TEMMINK et al. 1970). *Olpidium* species are obligate parasites of plant roots, and different isolates can exhibit host specificity, although the processes remain unclear. These chytrids use a 'resting spore' structure to overwinter, and in the spring motile zoospores germinate from sporangia to infect the roots of newly emerging plants (CAMPBELL 1996). Very low concentrations of *O. brassicae* zoospores can facilitate virus transmission, and virus infection of roots is more efficient with viruliferous zoospores than by mechanical inoculation. Causal associations linking virus transmission with fungal infection include occurrence of the infection foci of both the virus and *Olpidium* just behind the root tip in the region of cell elongation (FRY and CAMPBELL 1966; TEAKLE 1962). In addition, virus transmission is reduced drastically by treatments such as heating, drying, or chemicals that reduce the viability of zoospores (CAMPBELL and FRY 1966; KASSANIS 1962). Direct evidence supporting fungal transmission of TNV has been obtained in experiments showing that zoospores germinated from a single sporangium isolated from TNV-infected roots transmit virus, but zoospores from sporangia isolated from healthy roots are unable to facilitate infection (TEMMINK et al. 1970).

Several investigations have shown that the extent of TNV and STNV transmission depends on a number of interactions related to compatibility of the fungal isolates, the virus strains, and the host species. It appears that the ability of *Olpidium* strains to establish a productive fungal infection of the host per se is not correlated directly with vector efficiency or specificity of transmission. For example, tests of the ability of TNV strains A and D to be transmitted to different host plants by *Olpidium* showed that some of the fungal isolates that were highly pathogenic on certain hosts were unable to transmit virus, whereas other *Olpidium* isolates that were unable to establish productive infections transmitted virus efficiently (KAS-SANIS and MACFARLANE 1965; TEMMINK et al. 1970). Several studies using filtration and centrifugation also revealed that transmission is highly correlated with the ability of TNV and STNV to adsorb to the zoospore (CAMPBELL and FRY 1966; FRY and CAMPBELL 1966; KASSANIS and MACFARLANE 1964; TEAKLE 1960, 1962; TEMMINK et al. 1970). Taken together, the evidence suggests that adsorption of TNV to the fungus is a primary criterion for transmission, but that secondary factors influencing the specificity and efficiency of transmission are affected to a considerable extent by the fungal isolate, the host plant, and the environmental conditions during infection.

Electron microscopy has provided additional evidence bearing on the mechanism of TNV and STNV transmission by *Olpidium* (TEAKLE 1962; TEMMINK et al. 1970). In the first critical phase of dispersal, virus particles adsorb to the surface of germinating zoospores; the zoospore swims to the root and adheres at the zone of cell elongation. Several incompletely characterized infection stages then occur during the subsequent invasion of the root cells by the fungus. Time-course observations of infection show that the single polar flagellum initially is retracted into the cytoplasm of the zoospore, an event which presumably mediates transfer of adsorbed virus from the flagellum to the fungal protoplasm (CAMPBELL 1996). Next, while a fungal cyst forms in close association with the plant cell wall, a central channel connecting the fungal and host cell walls develops. Subsequently, the fungal protoplasm moves through this channel into the host cytoplasm (TEMMINK et al. 1970). At this stage of infection, TNV and STNV apparently come into intimate contact with the host cytoplasm, where they begin to multiply.

2.2 Panicovirus Satellites

A second group of satellite viruses is associated with panicum mosaic virus (PMV; Fig. 1). PMV is the type member of a new group that we have provisionally designated the *Panicovirus* genus (TURINA et al. 1998). PMV was first isolated from native grasses in Kansas in the 1950s, and its satellite virus (SPMV) was subsequently identified in the mid 1970s (BUZEN et al. 1984; NIBLETT and PAULSEN 1975). Several isolates of PMV have been obtained from the midwest and southern regions of the United States. These isolates have very narrow natural and experimental host ranges, which are primarily restricted to the Panicoid tribe of the *Gramineae* (ABU-SAMAH and HOLCOMB 1976; NIBLETT and PAULSEN 1975). PMV induces mild mosaic symptoms in pearl millet, but co-infections with SPMV generally result in severe mosaic, stunting, and reduced seed-set (MASUTA et al. 1987). Two serologically related isolates of PMV include St. Augustine decline virus (SADV) and centipede grass mosaic virus (CGMV). These viruses are common on diseased lawns in Texas, Louisiana, Arkansas, and South Carolina (HOLCOMB et al. 1989). Both SADV and CGMV support satellite viruses that are serologically related to SPMV, and these co-infections elicit disease symptoms that range from a mild mosaic to a lethal necrosis that can destroy entire lawns. No biological vector has yet been identified for any of the panicovirus isolates, but it is likely that spread of SADV and CGMV occurs primarily by mechanical transmission from infected plants during mowing (HAYGOOD and BARNETT 1992). However, it is not yet clear whether or not a soil-borne mode of transmission exists. CAMPBELL (1996) suggests that it would be worthwhile to explore strains of *Olpidium bornovanus* to determine whether these soil fungi can facilitate transmission of members of the *Tombusviridae* such as PMV.

PMV supports replication of serologically distinct SPMV isolates (BUZEN et al. 1984). In addition, Molinia streak virus (MSV), a virus isolated in Europe that is serologically related to PMV (PAUL et al. 1980), can act as a helper virus for replication of SPMV (BUZEN et al. 1984). Although selective activation of the PMV satellite viruses by specific helper virus isolates has not been demonstrated, not all serotype combinations have been tested; thus, some isolates may vary in their ability to support replication of various SPMV strains (BUZEN et al. 1984; HOL-COMB et al. 1989).

2.3 Maize White Line Mosaic Virus Satellite

Maize white line mosaic virus (MWLMV) is widely distributed in corn-growing areas of the northeastern and northcentral United States (LOUIE et al. 1982). MWLMV or a serologically related virus has also been observed in Italy and France (LOUIE et al. 1982). Symptoms caused by MWLMV vary depending on the time of the year and the age of the plants. Late in the season, corn plants frequently develop chlorotic white lines interspersed within the mosaic patterns, but a variety of mosaic and mottle phenotypes also appear in leaves during the early acute stages of infection. Under some circumstances, MWLMV can result in substantial yield losses in experimental plots. In heavily infected fields, losses can range up to 45% in field corn and to nearly 100% in sweet corn (LOUIE et al. 1982). Fortunately, extensive yield losses are not widespread because of the relatively low incidence of the virus in most commercial fields.

Several tests have confirmed that MWLMV and its satellite, SMWLMV, are transmitted readily through the soil; however, a specific vector has not yet been identified (LOUIE et al. 1982). Seed transmission has not been described, even though extensive tests show that virus can be detected in pollen, anthers, silks, and seeds of infected plants (LOUIE et al. 1982). MWLMV and SMWLMV normally are not mechanically transmissible to leaves, but high rates of infection can be obtained by making incisions with a sharp scalpel through a drop of dilute virus placed on corn seed or by puncturing the vascular tissue (LOUIE 1995; ZHANG et al. 1991a).

3 Physicochemical Properties of Helper Viruses

The properties of the helper viruses are mentioned only briefly here in order to provide a synopsis of their general properties. A more extensive consideration of their structure and genome organization follows near the end of the chapter in conjunction with a discussion of the helper virus determinants required for support of their respective satellite viruses. Other than TMV, helper viruses that support the replication of satellite viruses have superficially similar particle morphologies and physicochemical properties. The icosahedral virions of TNV, PMV, and MWLMV are approximately 28–30 nm with a T = 3 structure composed of 180 protein subunits. These viruses have small single-stranded RNA genomes that range in size from about 3800 to 4300 nucleotides (nt). Detailed sequence information has been accumulated for the helper viruses TNV-A (MEULEWAETER et al. 1990), TNV-D (COUTTS et al. 1991; MOLNAR et al. 1997), and PMV (TURINA et al. 1998). The complete nucleotide sequence has also been determined for satellite viruses associated with TNV, PMV, and MWLMV (BERGER et al. 1994; DANTHINNE et al. 1991; MASUTA et al. 1987; YSEBAERT et al. 1980; ZHANG et al. 1991a).

4 Structure of Satellite Virus Particles

The plant satellite viruses discovered to date all have similar physicochemical properties (Table 1). These include icosahedral virions of approximately 17 nm (Fig. 1) composed of protein subunits ranging in size from 17 to 24 kDa that encapsidate positive-sense ssRNA genomes varying from about 850 to 1250 nt that encode capsid (Table 1). The capsid proteins of the satellite virus particles are serologically distinct from those of their helper viruses, and their genomes have little nucleotide sequence relatedness as assessed by molecular hybridization analyses (MASUTA et al. 1987; SHOULDER et al. 1974; ZHANG et al. 1991a). The nucleotide sequences of STNV (DANTHINNE et al. 1991; YSEBAERT et al. 1980), SPMV (MASUTA et al. 1987; BERGER et al. 1994), and SMWLMV (ZHANG et al. 1991a)

Table 1. Structural comparisons of plant satellite viruses

Satellite virus	Genome (nt)	CP (amino acids)	CP (kDa)	Diameter (nm)
SPMV	826	157	17	16
STMV[a]	1059	159	17.5	18
SMWLMV	1168	208	24	–
STNV-1	1239	195	21	18
STNV-2	1245	197	21[b]	18

[a] For STMV details, see review by DODDS in this volume.
[b] Relative mobility in SDS-PAGE is approximately 24 kDa.
(– not determined)

have been determined, and their genome organization is depicted in Fig. 2. The genomes have some similarities in structure and coding capacity, but important individual differences are described below.

STNV, STMV, and SPMV have been subjected to extensive high-resolution structural analysis (BAN et al. 1995; BAN and McPHERSON 1995; JONES and LILJAS 1984; LARSON et al. 1993), but no definitive physical information is available concerning SMWLMV. Electron microscopy, solution scattering experiments with X-rays or neutrons, and X-ray crystallographic data have shown that the satellite viruses are composed of 60 identical protein subunits with a $T = 1$ icosahedral symmetry. X-ray crystallography reveals that STNV, STMV, and SPMV contain an eight-stranded 'jelly-roll' β-barrel that forms the core of each capsid protein subunit (BAN et al. 1995). However, important differences exist because SPMV has a 'jelly-roll' topology in the central portion of the CP, whereas STNV and STMV capsid proteins have a 'jelly-roll' structure at the carboxy-terminal region (BAN and McPHERSON 1995; BAN et al. 1995; JONES and LILJAS 1984; LARSON et al. 1993). BAN et al. (1995) also noted that five of the first 16 amino acids of the three satellite viruses are basic (lysine and/or arginine), and they suggested that this N-terminal region of the CP interacts with the satellite virus RNA during encapsidation. An additional unusual feature of SPMV is that its CP is used to encapsidate sat-RNAs (A.O. Jackson, unpublished observations). Both satellites are dependent on the PMV helper virus for replication. All other well-characterized sat-RNAs are encapsidated by their respective helper viruses, although one incompletely documented report suggests that a sat-RNA in a TNV infection is encapsidated by STNV (FRANCKI 1985; ROOSSINCK et al. 1992). It will be interesting to determine whether SPMV and its sat-RNAs share similar sequence motifs or structural elements which interact with the SPMV capsid protein.

5 Organization of Satellite Virus Genomes

5.1 STNV RNA

Early studies revealed that STNV-1 and STNV-2 are serologically distinct and have differences in the peptide profiles and amino acid compositions of their capsid proteins (REES et al. 1970). More recently, isolates of STNV-1 and STNV-2 have been extensively characterized, and substantial variations have been detected in their nucleotide sequence (DANTHINNE et al. 1991).

Sequence analyses of STNV-1 RNA revealed a genome of 1239 nt (YSEBAERT et al. 1980) that is illustrated schematically in Fig. 2. The 1245 nt RNA of STNV-2 has a similar genome organization but only 60% nucleotide sequence identity with STNV-1. The 5′ termini of STNV RNAs lack the m^7Gppp cap structure that is associated with most viral and messenger RNAs of eukaryotes. The untranslated leader sequences of STNV-1 and STNV-2 are 29 and 32 nt, respectively, and are identical for 21 of the first 27 nt from the 5' terminus. Downstream of the un-translated leader sequence, the STNV RNAs contain an open reading frame (ORF) which encodes the 21.5 kDa capsid protein (Fig. 2). The CP start codon context for STNV-1 (5'-AAC*A*U*G*G-3') is slightly better than that of STNV-2 (5'-AGA*A*U*G*A-3'). This is the only substantial ORF encoded on the STNV-1 RNA and, therefore, the genome appears to be monocistronic. STNV-2 RNA is similarly organized, although the predicted 21.5 kDa capsid protein ORF shares only 50% amino acid sequence similarity with STNV-1 (DANTHINNE et al. 1991). In addition, the STNV-2 capsid protein has an apparent molecular mass of 24 kDa, as estimated by gel electrophoresis of in vitro translation products of STNV-2 RNA and CP isolated from purified satellite virus particles (DANTHINNE et al. 1991). Sequence analyses of several cDNA clones derived from STNV RNA-2 populations have also revealed some sequence heterogeneity in the 3'-untranslated region of the genome (DANTHINNE et al. 1991).

The 3'-untranslated trailer regions of the STNV RNAs, representing the most conserved portion of the genome, are approximately 620 nt. The predicted sec-ondary structures of these 3'-untranslated trailer regions indicate nearly identical overlapping conformations (DANTHINNE et al. 1991). This high degree of secondary structure is reflected physically by resistance to thermal denaturation and nuclease degradation. The highly base paired property may account in part for the ability of STNV RNA to survive for extended periods in vitro and in inoculated leaves under conditions which degrade the RNAs of other viruses (MOSSOP and FRANCKI 1979). However, this portion of the genome undoubtedly has very critical biological roles related to its recognition by the helper virus replicase, its capsid protein associa-tions, and interactions with host components. Although the 3'-terminal portion of the STNV genome can theoretically be folded into a tRNA-like structure similar to methionine tRNA (YSEBAERT et al. 1980), thus far there is no evidence that the RNA is aminoacylated during the infection process. However, as described below, the 3' end of STNV RNA does appear to be involved in cap-independent trans-lation.

5.2 SPMV RNA

The 826-nt genome of SPMV is the smallest satellite virus RNA described to date (MASUTA et al. 1987). Several isolates of SPMV and their helper viruses have been reported in centipede grass and St. Augustine grass infections (HOLCOMB 1974; HOLCOMB et al. 1989). The general structure and organization of the genome is illustrated in Figs. 1 and 2. Characterization of the SAD strain (BERGER et al. 1994) revealed no significant differences from the Kansas isolate sequenced by MASUTA et al. (1987).

Like STNV, the 5' nucleotide of SPMV appears to be phosphorylated rather than capped, although the 5' nucleotide of the former has an adenine instead of the guanine residue of SPMV RNA (MASUTA et al. 1987). However, the 88 nt SPMV nontranslated 5' leader sequence is nearly three times the length of the analogous

STNV region. Both SPMV and STNV RNAs have the potential to form weakly energetic hairpin structures slightly upstream of their CP initiation codons, which may have a role in their efficient translation in vitro. The sequence surrounding the SPMV CP start codon (5'-CUGAUGG-3') has a permissive context for initiation of translation (Lutcke et al. 1987). The CP ORF encodes a 157 amino acid protein of 17 kDa that is nearly identical to the amino acid composition determined directly from purified SPMV capsid protein (Buzen et al. 1984; Masuta et al. 1987). Preliminary investigations using site-directed oligonucleotide mutagenesis to disrupt the predicted SPMV ORFs have confirmed that ORF1 encodes the CP (M. Turina and K.-B.G. Scholthof, unpublished results).

Downstream of the CP ORF, SPMV RNA encodes a 6.3 kDa ORF (ORF2) of uncertain function (Fig. 2). Two coding regions, ORF3 and ORF4, on the negative strand (not illustrated) potentially could express polypeptides of 7.1 kDa and 11 kDa, respectively (Masuta et al. 1987). STMV is the only other satellite virus that encodes more than one ORF (Mirkov et al. 1989), but the biological role of the 6.8 kDa encoding ORF (Fig. 2), which precedes the CP ORF, is unknown because its disruption does not affect the replication or movement of the satellite virus in tobacco (see Dodds review, this volume) (Routh et al. 1995).

The 5'- and 3'-terminal nucleotides of SPMV and its helper virus have some limited sequence similarities. The seven 5'-proximal nucleotides of PMV and SPMV are identical (5'-GGGUAUU-3'). In contrast, only three of the 3'-terminal nucleotides are identical, although introducing a gap of one nucleotide and a single base mismatch substantially improves the alignment between the SPMV (5'-CUAG-GACCC-3') and PMV (5'-CCAGG-CCC-3') 3' termini (Turina et al. 1998). No other obvious stretches of homology have been detected between PMV and SPMV. The 3'-terminal 267 nt of SPMV RNA, which includes the putative ORF2, is predicted to form stable base-paired regions (Masuta et al. 1987). It is anticipated that the uncharacterized secondary structure may contain important recognition signals for the helper virus replicase, and the structure may have as yet undefined roles in regulation of replication and/or translation, and may also provide sites for initiation of encapsidation.

Beyond these similarities in relatedness of the terminal nucleotides of SPMV and PMV and those of other satellite-helper combinations, very little definitive information is available about the evolution of satellite viruses. However, one possible clue to evolution of the SPMV CP comes from a sat-RNA of bamboo mosaic potexvirus (BaMV), which encodes a 20 kDa nonstructural protein with 46% identity to the 17 kDa SPMV CP (Liu and Lin 1995). This protein is expressed from the sat-RNA during plant and protoplast co-infections with BaMV (Lin et al. 1996). However, the BaMV sat-RNA is encapsidated by the rod-shaped helper virus CP to form 'rodlets', and the 20 kDa protein is not obviously associated with these structures. Although the BaMV protein appears to be dispensable for replication and for cell-to-cell movement, it may act as an RNA-binding protein to enhance sat-RNA movement in plants (Lin et al. 1996). Therefore, it is plausible that the 20 kDa protein may have evolved from a recombination of the sat-RNA progenitor with a conserved host gene encoding an RNA-binding protein. This notion forms the basis for a model whereby an ORF from a common host was incorporated via separate recombination events into both the BaMV sat-RNA and an SPMV RNA precursor. Subsequently, the SPMV progenitor could have undergone a series of mutations necessary for evolution into a capsid protein with specificity for encapsidation of SPMV RNA.

5.3 MWLMV Satellite Virus RNA

The 1168 nt genome of SMWLMV has some features in common with those of other satellite viruses (Table 1; Fig. 2). For example, SMWLMV RNA appears to lack a 5'-terminal cap structure, and the 3' terminus ends in 'CCC'. From the limited helper virus sequence that is available, MWLMV and its satellite virus share identity at only two 'C' residues at the 3' terminus (ZHANG et al. 1991a). Further characterization of the helper virus will clarify its relatedness with the satellite virus RNA, but based on nucleic acid hybridization experiments and the available sequence information, little appreciable homology exists between SMWLMV and its helper virus (ZHANG et al. 1991a).

A 160 nt untranslated leader sequence at the 5' end of the SMWLMV genome precedes a 24 kDa ORF whose high expression is facilitated by a good translational initiation context (5'-GUC*AUG*G'-3). In vitro translation of this satellite RNA results in the accumulation of a 24 kDa protein that can be immunoprecipitated with antiserum derived from the satellite virus particle (GINGERY and LOUIE 1985; ZHANG et al. 1991a). No additional ORFs of appreciable size are encoded by SMWLMV RNA, and a 351 nt untranslated trailer sequence follows the CP stop codon. The secondary structure of the 3' terminus and its biochemical or biological significance have not been analyzed.

6 Replicative Forms of Satellite Viruses

The helper viruses associated with STNV, SPMV, and SMWLMV are predicted to replicate through minus-sense or double-stranded RNA intermediates. The plus-sense ssRNA genome is copied by the helper virus-encoded replicase directly into a minus-sense (complementary) RNA that may function as a dsRNA intermediate template for asymmetric transcription of progeny plus-stranded RNAs. Thus far, there is no evidence to suggest that the satellite viruses deviate from this replicative strategy to produce rolling circle intermediates, as is the case for some satellite RNAs (see the chapter by Symons). Moreover, helper viruses support the replication of the satellite viruses, implying that the replicase proteins function in *trans* during amplification of the satellite virus RNAs, although this does not necessarily mean that intimate associations do not occur between the replicating satellite and helper RNAs. Such associations are likely, since the satellite virus RNA must compete for limited amounts of the helper component replicase to facilitate preferential amplification of the satellite virus RNA. These interactions thus have important implications for understanding the mechanisms of action of replicase proteins of the helper viruses, and the restricted co-evolution of sat-RNAs and satellite viruses within particular virus families.

It is unlikely that subgenomic mRNAs are synthesized during replication of STNV, since only a single major ORF exists on the genome. However, since SPMV has an ORF downstream of the CP ORF in a relatively poor translational start context (5'-UGG*A*UGU-3'), it is possible that an sgRNA may be necessary for the expression of the putative 6.3 kDa protein (Fig. 2). Interestingly, the 5'-proximal region of the minus-sense RNA of SPMV has a favorable translational start context (5'-AAA*A*UGG-3') preceding ORF3 that could express a 7.1 kDa protein (Masuta et al. 1987). We do not yet know whether proteins other than the CP are expressed, and mutational and biochemical analyses are underway to determine if ORFs 2, 3, or 4 are functional, and if they have essential roles in the life cycle of SPMV.

7 Cap-Independent Translation

Important similarities among the satellite viruses and their helper viruses are the lack of 5'-terminal cap structures and the polyadenylated 3' termini that are usually associated with translationally competent eukaryotic mRNAs. The small RNA helper viruses and their associated satellite viruses replicate to very high levels in the cytoplasm, so their nonconventional RNAs presumably have evolved to evade the host processes that target uncapped and nonpolyadenlyated RNAs for rapid cellular turnover. The nature of these putative evasion strategies is not well understood at any level, yet it is evident that the satellite virus RNAs are protected from host degradation while undergoing remarkably efficient replication, translation, and invasion of the host plant. STNV and SPMV have distinct nontranslated leaders of 21 and 89 nt, respectively. Computer analyses indicate that both RNAs have weakly energetic hairpin loops at the 5' termini near a short sequence that is complementary to the 18 S rRNA that presumably targets binding of the 40 S ribosome complex. Beyond this, very little experimental information is yet available concerning the mechanisms leading to efficient translation of SPMV RNAs. However, the 5' and 3' termini of STNV have been studied in some detail, and the studies described below present intriguing features that may be important for RNA stability and translation of the capsid protein (Danthinne et al. 1991, 1993; Jackson et al. 1995; Timmer et al. 1993).

The lack of a cap structure normally abrogates efficient translation of cellular mRNAs, although notable examples of 'cap-independent' translation mechanisms are known for a number of viral RNAs. This feature implies that these viral RNAs must compete effectively with capped mRNAs to sequester translation initiation components in order to facilitate preferential translation of uncapped viral RNAs over those of the capped host mRNAs (IIZUKA et al. 1995).

STNV RNA appears to be an authentic cap-independent message, because addition of a cap fails to enhance the formation of initiation complexes under translation conditions in vitro, or to increase the rate of CP synthesis over that of uncapped STNV RNA (DANTHINNE et al. 1993; TIMMER et al. 1993). In addition, chimeric mRNAs with exchanges of the 5'- and 3'-untranslated regions of STNV RNA and an α-globin mRNA in various combinations have revealed that interactions between the STNV 5'- and 3'-untranslated regions are essential for maintenance of cap-independent translation (TIMMER et al. 1993). For example, exchange of the 5'-untranslated leader of the α-globin mRNA for the STNV RNA leader does not convert globin mRNA into a cap-independent mRNA (TIMMER et al. 1993).

Specific regions within the 3'-untranslated STNV RNA trailer sequence also facilitate efficient translation of the CP in wheat germ in vitro translation systems. Progressive deletion analyses of the 3'-untranslated trailer of the STNV have resulted in identification of a putative 150 nt translational enhancer region that is critical for CP expression and that has a protective role in RNA stability (DANTHINNE et al. 1993; TIMMER et al. 1993). This region appears not to be a spacer sequence per se, because addition of an unrelated sequence to restore the length of the trailer failed to yield compensatory wild-type levels of translation (TIMMER et al. 1993). However, strong interactions are evident between the 5'- and 3'-untranslated regions, since capping of the truncated RNA can restore its translational efficiency. Interestingly, sequence comparisons and alignments with the STNV enhancer-like element suggest that SPMV RNA may also have a 3'-translational enhancer domain (K.-B.G. Scholthof, unpublished observations), which could be important in facilitating cap-independent initiation of translation.

8 Helper Virus Requirements for Replication of Satellite Viruses

8.1 Tobacco Necrosis Virus

More than 20 European isolates of TNV have been described that can be classified into the TNV-A and TNV-D groups defined by Kassanis (BABOS and KASSANIS 1963b). These isolates are typed according to their serological relatedness, to their symptom phenotypes on French bean and tobacco, to *Olpidium brassicae* transmission, and to their differential ability to support the replication of specific STNV strains. TNV-A supports both STNV-1 and STNV-2 replication, but TNV-D supports only STNV-C (BABOS and KASSANIS 1963a; MEULEWAETER et al. 1993; UYEMOTO et al. 1968).

The single-stranded TNV RNA is approximately 3.8 kb, with a genome organization similar to that of the carmoviruses (Fig. 3), but TNV-A and TNV-D have sufficient differences to merit a separate classification as necroviruses (COUTTS et al. 1991; MEULEWAETER et al. 1990; MOLNAR et al. 1997). As is the case for all viruses within the *Tombusviridae*, the replicase is composed of two proteins (Fig. 3). The smaller, approximately 22 kDa, and more abundant of these two proteins is translated by initiation at the 5′-proximal ORF. The larger 82 kDa protein is synthesized in minor amounts using a translational readthrough strategy of an amber codon, as is characteristic of viruses in the *Tombusviridae*. The amino acid sequences of the TNV-A and TNV-D replicase proteins are 45% identical (COUTTS et al. 1991; MOLNAR et al. 1997), but the predicted movement proteins of TNV-A and TNV-D appear to have evolved independently, because they share little direct amino acid sequence similarity. In contrast, the capsid proteins of TNV-A and TNV-D are more closely related and show about 57% identity within the shell domain. The TNV capsid proteins have more similarity to the CP of southern bean mosaic sobemovirus than to those of the carmoviruses (COUTTS et al. 1991). These results provide a persuasive argument that TNV-A and TNV-D have evolved separately via molecular recombination from diverse components, some of which may be of cellular origin. The construction of an infectious cDNA clone of TNV-A, in combination with the recently reported infectious clone of TNV-D (MOLNAR et al. 1997), could provide a unique opportunity to develop chimeric infectious transcripts that would be useful for elucidating the domain(s) required for satellite virus replication and movement (COUTTS et al. 1991).

8.2 Panicum Mosaic Virus

PMV virions encapsidate a single stranded RNA genome of 4326nt that potentially can encode seven proteins (Fig. 3). Sequence analyses reveal that PMV has similarities to the carmo-, necro-, and machlomoviruses, but the virus is sufficiently unique to assign it to a separate group, which we have provisionally designated the *Panicovirus* genus (TURINA et al. 1998). As with TNV, the replicase proteins are produced by the use of an amber suppressor codon to encode a 48 kDa protein and minor amounts of a 112 kDa readthrough protein (Fig. 3). The infectivity of transcripts derived from deletions of an infectious cDNA clone reveal that p48 and p112 are sufficient for the replication of PMV and SPMV in pearl millet protoplasts (M. Turina and K.-B.G. Scholthof, unpublished observations). Thus far, only one subgenomic RNA (sgRNA) has been detected in association with PMV infections (TURINA et al. 1998). This sgRNA has five potential ORFs, including a 3′-proximal CP, and preliminary results indicate that each of these ORFs is necessary for spread of the virus in plants (M. Turina and K.-B.G. Scholthof, unpublished observations).

8.3 Maize White Line Mosaic Virus

MWLMV has 30 nm icosahedral particles that are composed of a single-stranded positive-sense RNA of approximately 4200 nt and a capsid protein of approximately 35 kDa (ZHANG et al. 1991b). However, two dsRNA species are present in preparations from infected corn leaves (GINGERY and LOUIE 1985; ZHANG et al. 1991b). Subgenomic dsRNAs are frequently observed in infections of small icosahedral viruses, and these appear to be related to the subgenomic mRNAs. Thus, the smaller dsRNA is of a sufficient size to correspond to a subgenomic mRNA of approximately 2 kb that may be expressed during replication. Preliminary characterizations (ZHANG et al. 1991b) indicate that MWLMV may be related to the 'small RNA viruses' in the carmo-, necro-, and machlomovirus groups within the *Tombusviridae*, but the complete sequence needs to be evaluated to verify this possibility. MWLMV and its satellite virus can both be transmitted to healthy sweet corn by planting seeds in soils that formerly contained MWLMV-infected plants. Although severe disease symptoms may occur, plants from infested soil often cannot be distinguished from healthy plants, and frequently no differences in symptoms are observed between plants infected with MWLMV alone and those co-infected with the satellite virus (GINGERY and LOUIE 1985). These findings are reminiscent of the TNV and STNV complexity, and they raise the possibility that a soil-borne agent, possibly a fungus, is a vector in the field (GINGERY and LOUIE 1985).

9 Replication of Satellite Virus Genomes

The interactions between satellite viruses and their helper viruses are poorly understood. This is partially because of the complexity observed in mixed infections of different strains of the helper and satellite viruses, restrictions imposed by their hosts, and uncharacterized environmental effects. To study the requirements for replication, ANDRIESSEN et al. (1995) used a replicase cassette of TNV-A to express the p23 and p82 replicase subunits (Fig. 3) from a cauliflower mosaic virus 35 S promoter, either transiently in protoplasts or from transformed tobacco. The presence of these two replicase proteins was sufficient to support STNV-2 replication (ANDRIESSEN et al. 1995), although the accumulation of STNV-2 RNA was only about 1% of that observed when wild-type TNV-A RNA and STNV-2 were co-electroporated into tobacco protoplasts. These experiments thus confirm that the helper virus replicase proteins can support replication of satellite viruses in *trans*. However, the low levels of expression of the replicase proteins, their relative ratios in transgenic plants, or perhaps other TNV-A encoded genes may have affected the accumulation of the satellite RNA.

Other experiments had previously indicated that the STNV CP and/or *cis*-elements within the CP ORF are necessary to maintain wild-type levels of RNA accumulation in protoplasts and plants (VAN EMMELO et al. 1987). For example, insertion of a 14 bp linker in the CP ORF of STNV resulted in a shift from accumulation of uniform mixtures of dsRNA and plus-sense RNA to primarily dsRNA. These results thus raised the possibility that the STNV CP has a regulatory role in replication, or that its coding region contains important *cis*-elements that affect the stability of plus-sense STNV RNA. An important future challenge is to define these elmements and to determine whether similar results can be obtained with other satellite viruses.

Recent experiments also indicate that the p48 and p112 PMV replicase subunits are sufficient to support the replication of the SPMV transcripts in protoplasts (M. Turina and K-B.G. Scholthof, unpublished observations). As is the case with TNV, the PMV experiments also suggest that satellite or helper virus-encoded proteins may act in concert with replicase subunits to regulate replication and accumulation of the satellite virus RNA. We are presently investigating these possibilities.

10 Conclusions and Future Prospects

The satellite viruses comprise a small group of self-encapsidating RNAs that interact closely with their host plants and helper viruses. Both the satellite viruses and the helper viruses appear to share some common evolutionary features. The helper viruses TNV-A, TNV-D, and PMV have similar genome organizations, particle structures, and replication strategies (Fig. 3). There are also considerable similarities in the four satellite viruses (Table 1; Fig. 2). Intriguing hints suggest that the satellite viruses and their helpers have a number of similar biological properties. For example, it is possible that fungal vectors or other soil inhabitants may have some role in the ecology of all of these agents. Therefore, it may be profitable to extend some of the classical studies performed on TNV and STNV to more thoroughly evaluate the extent and nature of soil transmission of the satellite virus complexes.

Whether the satellite viruses are merely 'molecular parasites' or have some beneficial role in the biology of helper viruses has not been explored in ecological studies. However, it is possible that the attenuating effects of the satellite viruses on the symptom phenotype may enhance host survival and hence provide a more densely distributed reservoir for maintenance of the helper virus. All the helper viruses have a relatively narrow host range, and, as KASSANIS (1962) originally suggested, the satellite virus may alter the metabolism of host cells so that they become more tolerant to replication of the helper virus. Thus, a long-term goal is to unravel the biological complexities of the helper-satellite-host interactions and their roles in ecology and pathology.

Detailed molecular genetic analyses need to be focused on several aspects of the satellite-helper virus interaction. Since the satellites viruses share similar strategies for replication and encapsidation, they provide unique tools for comparisons of specific host and helper virus RNA:protein interactions. The reagents necessary to obtain such information with TNV and STNV have been available for some time. Moreover, the availability of high-resolution crystallographic structural data on SPMV (BAN and McPHERSON 1995) and recent construction of infectious full-length cDNA clones of SPMV and PMV (TURINA et al. 1998) now provide an opportunity to investigate a large number of interactions with this helper satellite system. For example, we can now address the mechanisms of replication and packaging of satellite viruses, the strategies used to facilitate local and vascular movement in plants, and the helper-satellite interactions leading to disease exacerbation or attenuation of disease phenotypes. Specialized biological properties, such as identification of the determinants affecting the ability of the TNV helper and satellite to be transmitted by their fungal vectors, can also be explored in detail using modern molecular genetic techniques. Similar approaches can be used to investigate the elements involved in recognition and encapsidation of the sat-RNAs of PMV in SPMV virions.

In addition to fundamental advances, there may be some potential biotechnological applications, whereby we may harness satellite viruses as vectors for efficient expression of foreign genes (SCHOLTHOF et al. 1996). Vectors with reporter gene substitutions for the CP genes might also be exploited for quantative studies of replication, aspects of gene regulation, and visualization of the spread of satellite viruses. The feasibility of these approaches has been suggested by the work on satBaMV that has permitted replacement of the 20 kDa protein with the CAT gene (LIN et al. 1996). Therefore, numerous opportunities exist for future expansion of our knowledge about the biology and biochemical interactions occurring during replication and disease development by these rare and interesting helper-satellite virus systems.

Acknowledgements. We appreciate the helpful comments of HERMAN SCHOLTHOF and MASSIMO TURINA.

References

Abu-Samah N, Holcomb GE (1976) New hosts of St. Augustine decline virus. Phytopathology 66:215-216

Andriessen M, Meulewaeter F, Cornelissen M (1995) Expression of tobacco necrosis virus open reading frames 1 and 2 is sufficient for the replication of satellite tobacco necrosis virus. Virology 212:222-224

Babos P, Kassanis B (1963a) The behaviour of some tobacco necrosis virus strains in plants. Virology 20:498-506

Babos P, Kassanis B (1963b) Serological relationships and some properties of tobacco necrosis virus strains. J Gen Microbiol 32:135-144

Ban N (1995) The structure of satellite panicum mosaic virus at 1.9 A resolution. Nature Struct Biol 2:882-890

Ban N, Larson SB, McPherson A (1995) Structural comparison of the plant satellite viruses. Virology 214:571-583

Berger PH, Shiel PJ, Gunasinghe U (1994) The nucleotide sequence of satellite St. Augustine decline virus. Mol Plant Microbe Interact 7:313–316

Buzen FG, Niblett CL, Hooper GR, Hubbard J, Newman MA (1984) Further characterization of panicum mosaic virus and its associated satellite virus. Phytopathology 74:313 318

Campbell RN (1996) Fungal transmission of plant viruses. Annu Rev Phytopathol 34:87 108

Campbell RN, Fry PR (1966) The nature of the associations between Olpidium brassicae and lettuce big-vein and tobacco necrosis viruses. Virology 29:222–233

Coutts RHA, Rigden JE, Slabas AR, Lomonossoff GP, Wise PJ (1991) The complete nucleotide sequence of tobacco necrosis virus strain D. J Gen Virol 72:1521–1529

Danthinne X, Seurinck J, van Montagu M, Pleij CWA, van Emmelo J (1991) Structural similarities between the RNAs of two satellites of tobacco necrosis virus. Virology 185:605 614

Danthinne X, Seurinck J, Meulewaeter F, van Montagu M, Cornelissen M (1993) The 3′ untranslated region of satellite tobacco necrosis virus RNA stimulates translation in vitro. Mol Cell Biol 13:3340–3349

Francki RIB (1985) Plant virus satellites. Annu Rev Microbiol 39:151–174

Fry PR, Campbell RN (1966) Transmission of a tobacco necrosis virus by Olpidium brassicae. Virology 30:517–527

Gingery RE, Louie R (1985) A satellite-like virus particle associated with maize white line mosaic virus. Phytopathology 75:870–874

Haygood RA, Barnett OW (1992) Widespread occurrence of centipede grass mosaic in South Carolina. Plant Dis 76:46–49

Holcomb GE (1974) Serological strains of panicum mosaic virus. Proc Am Phytopathol Soc 1:21

Holcomb GE, Liu TZ, Derrick KS (1989) Comparison of isolates of panicum mosaic virus from St. Augustine grass and centipede grass. Plant Dis 73:355–358

Iizuka N, Chen C, Yang Q, Johannes G, Sarnow P (1995) Cap-independent translation and internal initiation of translation in eukaryotic cellular mRNA molecules. In: Sarnow P (ed) Cap-independent translation. Springer, Berlin Heidelberg New York, pp 155–177 (Current topics in microbiology and immunology, vol 203)

Jackson RL, Hunt SL, Reynolds JE, Kaminski A (1995) Cap-dependent and cap-independent translation: operational distinctions and mechanistic interpretations. In: Sarnow P (ed) Cap-independent translation. Springer, Berlin Heidelberg New York, pp 1 29 (Current topics in microbiology and immunology, vol 203)

Jones TA, Liljas L (1984) Structure of satellite tobacco necrosis virus after crystallographic refinement at 2.5 A resolution. J Mol Biol 177:735–767

Kassanis B (1962) Properties and behaviour of a virus depending for its multiplication on another. J Gen Microbiol 27:477–488

Kassanis B (1981) Portraits of viruses: tobacco necrosis virus and its satellite virus. Intervirology 15:57 70

Kassanis B, Macfarlane I (1964) Transmission of tobacco necrosis virus by zoospores of Olpidium brassicae. J Gen Microbiol 36:79–93

Kassanis B, Macfarlane I (1965) Interaction of virus strain, fungus isolate and host species in the transmission of tobacco necrosis virus. Virology 26:603 612

Kassanis B, Macfarlane I (1968) The transmission of satellite viruses of tobacco necrosis virus by Olpidium brassicae. J Gen Virol 3:227–232

Kassanis B, Nixon HL (1960) Activation of one plant virus by another. Nature 187:713–714

Kassanis B, Phillips MP (1970) Serological relationship of strains of tobacco necrosis virus and their ability to activate strains of satellite virus. J Gen Virol 9:119–126

Larson SB, Koszelak S, Day J, Greenwood A, Dodds JA, McPherson A (1993) Three-dimensional structure of satellite tobacco mosaic virus at 2.9 A resolution. J Mol Biol 231:375 391

Lin N-S, Lee Y-S, Lin B-Y, Lee C-W, Hsu Y-H (1996) The open reading frame of bamboo mosaic potexvirus satellite RNA is not essential for its replication and can be replaced with a bacterial gene. Proc Natl Acad Sci USA 93:3138 3142

Liu JS, Lin NS (1995) Satellite RNA associated with bamboo mosaic potexvirus shares similarity with satellites associated with sobemoviruses. Arch Virol 140:1511–1514

Louie R (1995) Vascular puncture of maize kernels for the mechanical transmission of maize white line mosaic virus and other viruses of maize. Phytopathology 85:139 143

Louie R, Gordon DT, Knoke JK, Gingery RE, Bradfute OE, Lipps PE (1982) Maize white line mosaic virus in Ohio. Plant Dis 66:167–170

Lutcke HA, Chow KC, Mickel FS, Moss KA, Kern HF, Scheele GA (1987) Selection of AUG initiation codons differs in plants and animals. EMBO J 6:43 48

Masuta C, Zuidema D, Hunter BG, Heaton LA, Sopher DS, Jackson AO (1987) Analysis of the genome of satellite panicum mosaic virus. Virology 159:329–338

Meulewaeter F, Seurinck J, van Emmelo J (1990) Genome structure of tobacco necrosis virus strain A. Virology 177:699–709

Meulewaeter F, Danthinne X, Coutts R, van Emmelo J (1993) Specificity of satellite activation by tobacco necrosis virus correlates with nucleic acid hybridization pattern between helper virus isolates. Virology 193:971–973

Mirkov TE, Mathews DM, Du Plessis DH, Dodds JA (1989) Nucleotide sequence and translation of satellite tobacco mosaic virus RNA. Virology 170:139–146

Molnar A, Havelda Z, Dalmay T, Szutorisz H, Burgyan J (1997) Complete nucleotide sequence of tobacco necrosis virus strain D^H and genes required for RNA replication and virus movement. J Gen Virol 78:1235–1239

Mossop DW, Francki RIB (1979) The stability of satellite viral RNAs in vivo and in vitro. Virology 94:243–253

Niblett CL, Paulsen AQ (1975) Purification and further characterization of panicum mosaic virus. Phytopathology 65:1157–1160

Paul HL, Querfurth G, Huth W (1980) Serological studies on the relationships of some isometric viruses of Gramineae. J Gen Virol 47:67–77

Rees MW, Short MN, Kassanis B (1970) The amino acid composition, antigenicity, and other characteristics of the satellite viruses of tobacco necrosis virus. Virology 40:448–461

Roossinck MJ, Sleat D, Palukaitis P (1992) Satellite RNAs of plant viruses: structures and biological effects. Microbiol Rev 56:265–279

Routh G, Dodds JA, Fitzmaurice L, Mirkov TE (1995) Characterization of deletion and frameshift mutants of satellite tobacco mosaic virus. Virology 212:121–127

Scholthof HB, Scholthof K-BG, Jackson AO (1996) Plant virus gene vectors for transient expression of foreign proteins in plants. Annu Rev Phytopathol 34:299–323

Shoulder A, Darby G, Minson T (1974) RNA-RNA hybridization using ^{125}I-labeled RNA from tobacco necrosis virus and its satellite. Nature 251:733–735

Teakle DS (1960) Association of Olpidium brassicae and tobacco necrosis virus. Nature 188:431–432

Teakle DS (1962) Transmission of tobacco necrosis virus by a fungus, Olpidium brassicae. Virology 18:224–231

Temmink JHM, Campbell RN, Smith PR (1970) Specificity and site of in vitro acquisition of tobacco necrosis virus by zoospores of Olpidium brassicae. J Gen Virol 9:201–213

Timmer RT, Benkowski LA, Schodin D, Lax SR, Metz AM, Ravel JM, Browning KS (1993) The 5′ and 3′ untranslated regions of satellite tobacco necrosis virus RNA affect translational efficiency and dependence on 5′ cap structure. J Biol Chem 268:9504–9510

Turina M, Maruoka M, Monis J, Jackson AO, Scholthof K-BG (1998) Nucleotide sequence and infectivity of a full-length cDNA clone of panicum mosaic virus. Virology 241:141–155

Uyemoto JK, Grogan RG, Wakeman JR (1968) Selective activation of satellite virus strains by strains of tobacco necrosis virus. Virology 34:410–418

Valverde RA, Dodds JA (1986) Evidence for a satellite RNA associated naturally with the U5 strain and experimentally with the U1 strain of tobacco mosaic virus. J Gen Virol 67:1875–1884

van Emmelo J, Ameloot P, Fiers W (1987) Expression in plants of the cloned satellite tobacco necrosis virus genome and of derived insertion mutants. Virology 157:480–487

Ysebaert M, van Emmelo J, Fiers W (1980) Total nucleotide sequence of a nearly full-size DNA copy of satellite necrosis virus RNA. J Mol Biol 143:273–287

Zhang L, Zitter TA, Palukaitis P (1991a) Helper virus-dependent replication, nucleotide sequence and genome organization of the satellite virus of maize white line mosaic virus. Virology 180:467–473

Zhang L, Zitter TA, Palukaitis P (1991b) Maize white line mosaic virus double-stranded RNA, replicative structure, and in vitro translation product analysis. Phytopathology 81:1253–1257

Satellite Tobacco Mosaic Virus

J.A. Dodds

1 Introduction

Satellite viruses contain particles and genomes that have structural and genetic features which indicate a viral origin but have lost (or never had) the ability to replicate by themselves in susceptible plant hosts. They lack an identifiable gene for a replicase of their own and so, in order to replicate, they need to be a dependent partner in a mixed infection with another virus, usually referred to as the helper virus. The helper virus is presumed to supply the replicase function the satellite virus lacks, but it is not needed for particle formation, since the satellite virus codes for its own capsid protein.

A well-characterized satellite virus is the spherical satellite tobacco mosaic virus (STMV; Fig. 1), which was initially discovered in California in association with the rod-shaped tobacco mild green mosaic tobamovirus (TMGMV, also known as TMV-U5) in *Nicotiana glauca* (tree tobacco) (VALVERDE and DODDS 1986, 1987; VALVERDE et al. 1991). There is no serological relatedness between STMV and TMGMV. STMV can adapt to and replicate with other tobamoviruses including tobacco mosaic virus (TMV) in many hosts (VALVERDE et al. 1991). Viruses from other groups that have been tested cannot support STMV. The properties of STMV, its natural and experimental genetic diversity, and the present

Department of Plant Pathology, University of California, Riverside, CA 92521, USA

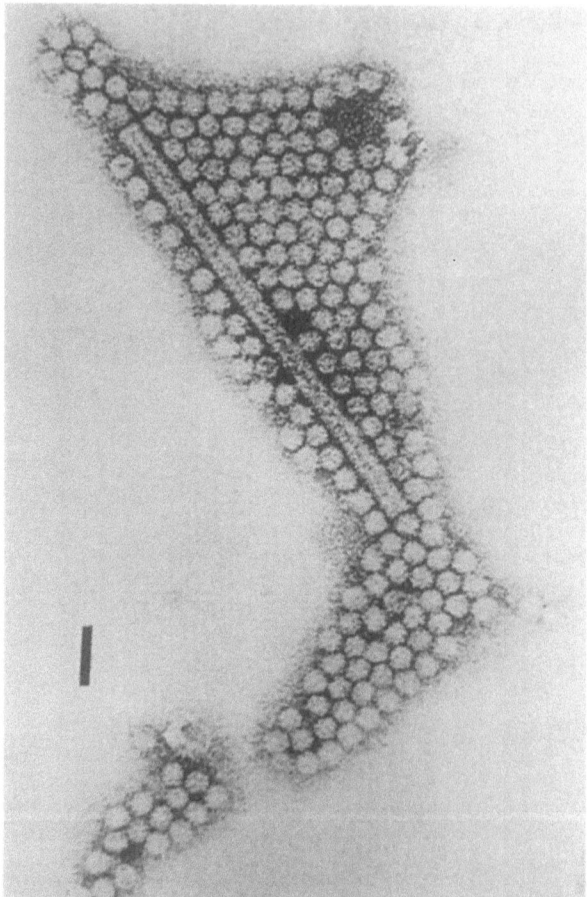

Fig. 1. Electron micrograph of a negatively stained purified preparation of TMGMV and STMV from doubly infected tobacco showing a full-length rod-shaped virion of TMGMV (300nm) surrounded by spherical virions of STMV (16–17nm). *Bar* represents 50nm. (From VALVERDE and DODDS 1987)

knowledge of the rules that condition its dependency on tobamoviruses are summarized in this chapter.

2 General Properties of STMV

The initial discovery of STMV was in association with TMGMV isolated from *N. glauca*. Some TMGMV isolates that had been biologically purified by passage through single local lesions on leaves of *N. sylvestris* (a local lesion host for TMGMV) had a prominent low-molecular-weight ($M_r = 0.6 \times 10^6$) dsRNA associated with them that was more abundant than the replicative form of TMGMV ($M_r = 4.3 \times 10^6$) (VALVERDE and DODDS 1986). This is now known to be the replicative form (RF) of the STMV RNA genome. Subsequent work has established that STMV exists as a spherical virus particle that is not infectious by itself but becomes so in the presence of TMGMV (VALVERDE and DODDS 1987). It is widespread in *N. glauca* throughout California (VALVERDE et al. 1991), as are TMGMV and several other viruses including cucumber mosaic cucumovirus and associated satellite RNAs, tobacco etch potyvirus, a carlavirus, and at least one additional uncharacterized virus (DODDS 1993). It is not uncommon for most or even all of these viruses, satellite viruses, and satellite RNAs to be present in the same tree tobacco plant, which grows as a tall well-branched shrub with small leaves, usually exhibiting prominent mosaic and yellowing symptoms of virus infection. TMGMV is also common in *N. glauca* in Spain, Israel, Australia, and other countries with Mediterranean climates, but there have been no additional reports of STMV. STMV has a wide host range and will infect several agronomic hosts experimentally, using a range of tobamoviruses as alternate helper viruses (VALVERDE et al. 1991; RODRIGUEZ-ALVARADO et al. 1994). These include tobacco, pepper, and tomato, but STMV has yet to be reported from a crop plant.

In controlled infections under greenhouse conditions STMV does not usually change the symptomatology expected of the helper tobamovirus. For example, TMGMV causes a mild mosaic in *N. glauca* and in *N. tabacum*, and STMV does not change this symptom. Consequently, TMGMV and STMV are unlikely to be the major contributors to the strong virus symptoms observed in *N. glauca* in the field. One clear exception to this generalization is the observed effects of STMV on the symptoms of tobamoviruses in peppers (RODRIGUEZ-ALVARADO et al. 1994; Fig. 2). In jalapeño and pimiento peppers the presence of STMV is associated with bright-yellow leaf patches not seen in single infections with tobamoviruses that infect pepper systemically. A severe leaf blistering in jalapeño that is induced by TMGMV is attenuated by STMV infection. Tobamovirus concentrations are suppressed by STMV in pepper compared with some other hosts in which symptoms are not affected, but the significance of this for the observed effects is unclear. In *N. glauca* and *N. tabacum* STMV also reaches high concentrations with only a slight negative effect on the high concentration of TMV or TMGMV.

STMV induces distinctive features in cells, including the accumulation of virus crystals and other protein bodies within unit membrane-bound structures (KIM et al. 1989). The membrane around these crystal inclusions is heavily vesiculated in a manner that suggests these are sites for genome replication. These STMV-specific features are found in cells that also have characteristic features associated with infection by TMGMV, confirming that individual cells are doubly infected. The replication of the two viruses does seem to be separately compartmentalized within a single cell, which has implications for how the satellite virus uses TMGMV gene products such as replicases. The dependency of STMV on tobamoviruses for replication has been established at the level of a single cell by mixed inoculation experiments in protoplasts (ROUTH et al. 1997; Fig. 3).

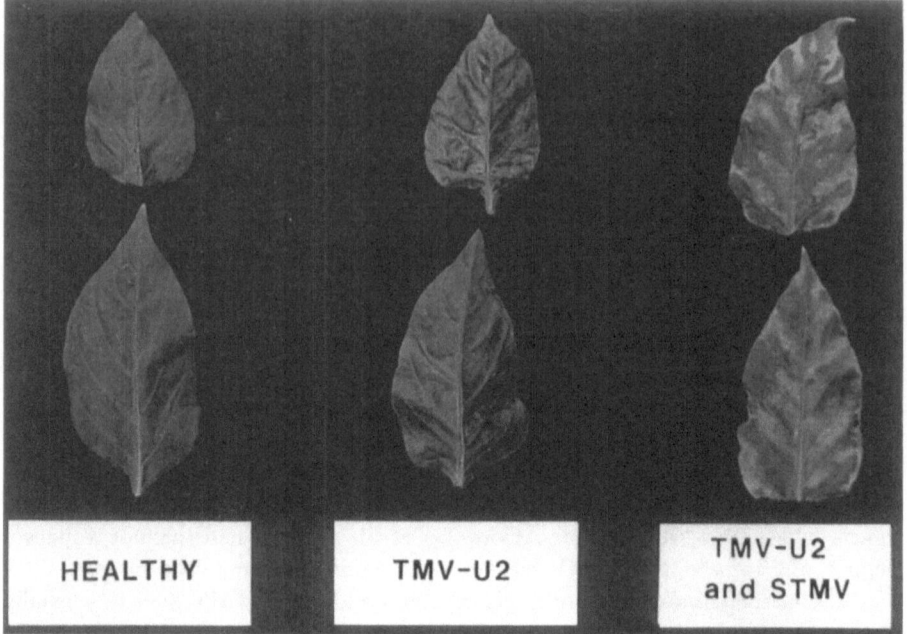

Fig. 2. Effect of TMGMV (labeled TMV-U2) infection and co-infection of TMGMV and STMV on pimiento pepper leaves from 6-week-infected plants. Light (yellow) patches are seen only in doubly infected leaves. (From RODRIGUEZ-ALVARADO et al. 1994)

3 Particle Structure and Genome Organization of STMV

STMV is a small (16–17nm in diameter) icosahedral particle consisting of 60 identical capsid protein subunits ($M_r = 17{,}000$ daltons), within which is a single molecule of a linear single-stranded RNA genome of 1059 nucleotides. The particle has been crystallized from purified preparations and the complete virion structure has been solved, including the resolution of up to 45% of the RNA within the particle (KOSZELAK et al. 1989; LARSON et al. 1993a,b). This RNA is in the form of 7-bp regions of duplex RNA, each associated with a set of capsid protein dimers which form a total of 30 internal cups within the particle where this duplex RNA can fit. The detection and description of the distribution of RNA inside the particle make STMV one of the most complete icosahedral viral structures known.

The complete nucleic acid sequence and organization of the STMV genome is known (MIRKOV et al. 1989). The 5′ terminal base was determined to be a non-phosphorylated adenosine residue, and there is a 52-base untranslated region (UTR). The first 19 bases have a clear resemblance to the corresponding sequence in the 5′ UTR of the genomes of cucumoviruses and bromoviruses, and a stem-loop structure common to these viruses is also predicted for some variants of STMV (YASSI and DODDS, unpublished). The first ORF codes for a 6.8-kDa protein which is a product obtained from in vitro translation experiments and has no sequence

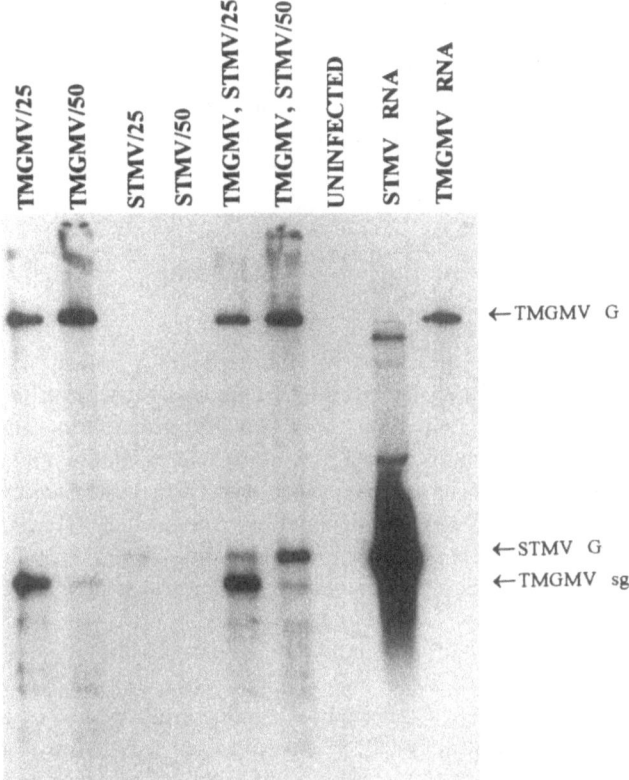

Fig. 3. Replication of STMV and TMGMV RNAs in protoplasts. Inoculations were with TMGMV alone, with STMV alone, or with both viruses, and protoplasts were analyzed 25 and 50h after inoculation. Controls were transcript RNAs used as inoculum. RNAs were detected with a mix of probes complementary to STMV and TMGMV genomic RNAs. STMV RNA accumulated in doubly inoculated protoplasts without any obvious effect on TMV RNA accumulation. STMV RNA inoculum degraded in singly inoculated protoplasts. (From Routh et al. 1997)

similarity with any other known plant viral gene. The second ORF starts at base 163 and encodes the capsid protein of STMV. It has a sequence that is dissimilar to that of other plant virus coat protein genes, and so no obvious taxonomic comparisons can be made. The 418-base 3′ UTR begins at base 641 and includes a complex region predicted to contain six pseudoknots. It is also predicted to assume a tRNA-like structure, terminates with GGCCCA, and can be aminoacylated with histidine (Felden et al. 1994; Gultyaev et al. 1994). The 3′ UTR is therefore similar to that of tobamoviruses and a clear sequence similarity between STMV and TMV has been detected (Mirkov et al. 1989). Most tobamoviruses have three pseudoknots in a 3′ UTR of only about 250 nucleotides. ORSV is an example of a tobamovirus that has a 3′ UTR about the same size as STMV, and also a more extensive set of pseudoknots than a typical tobamovirus (Gultyaev et al. 1994) The ssRNA genome of STMV is predicted to have a strong propensity to form internal secondary structures, including sequences that make up the open reading frames (KURATH et al. 1993b; LARSON et al. 1993b). A high degree of native secondary structure within particles has been confirmed, as discussed above. Secondary structures for the same RNA molecules when they are active in replication or translation have not been analyzed, other than the ones already mentioned in the 5' and 3' UTRs. Some of the main features of the STMV genome are summarized in Fig. 4.

4 Natural Variability of the STMV Genome

The genome of STMV is quite clearly organized like a typical virus, including untranslated regions, and an open reading frame for a structural protein (DODDS 1991). Predicted 5' UTR stem loops, and 3' UTR pseudoknots and tRNA-like structures complete this picture. The RNA can be encapsidated, and strong RNA-protein interactions have been observed, including the need to have repeated RNA secondary structures (dsRNA regions) within the particle. These genetic and structural elements are all included in 1059 nucleotides, and this small genome size makes STMV a good system for analyzing the natural genetic diversity of viral genomes.

Diversity was apparent after the initial detection of STMV by dsRNA analysis, and variants with different sized dsRNA RFs were described (VALVERDE and DODDS 1987). One of these was selected for cloning and sequencing, and it became the type strain (MIRKOV et al. 1989). This strain itself was found to be genetically heterogeneous, and was shown to exist as a typical quasispecies population of RNA molecules (KURATH et al. 1992). Forty-two independent full-length clones were placed into 16 groups, based on analysis by RNAse protection assays (RPA). There were two major populations, depending on the base at position 751 in the 3' UTR, which was either an A or a C. Genomes with this alternative were sufficiently abundant for both to be detected by RPA analysis of the entire population. Beyond this difference, microheterogeneity sites, determined by sequencing one clone from each of the 16 groups, were scattered at random over the genome with no preference for noncoding regions. cDNA clones from 13 of the 16 variants were infectious. The nature of the differences detected was not random, and a strong bias (62%) for substitution of an A for a G was recorded.

Infectious RNA from two full-length cDNA clones with only five single base differences, including the A or C at position 751 (which allows for convenient detection of one or the other by RPA), were tested for their ability to cross-protect or co-exist in experimental mixed infections (KURATH and DODDS 1994). When inoculation with the second variant took place more than 3 days after inoculation with the first variant, the first variant predominated in the plants for the duration of the experiment. When the inoculation delay was shorter, both variants tended to establish themselves, as is the case for the type strain isolate. This experiment indicates that minor variation in genome sequence is sufficient for viruses to exhibit the phenomenon known as cross-protection.

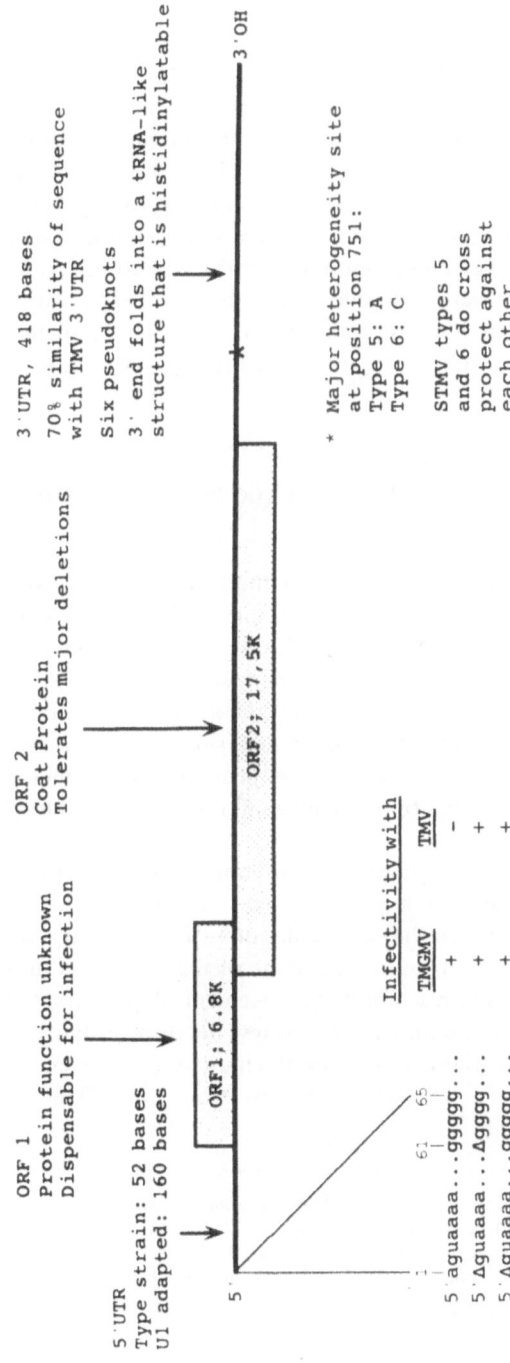

Fig. 4. Genome organization of STMV with a summary of some known features. (Prepared by M. Ngon A Yassi)

Fifteen independent isolates were compared with RPA, using probes that covered the entire genome, in order to describe the dominant variant present in each quasispecies (KURATH et al. 1993a; Fig. 5). They were collected within a 20-km^2 area which included the experimental station at UCR, where *N. glauca* is common. Ten distinct genomes were detected, and regions of the genome were very divergent from the type strain in some of them. More divergence was encountered in the 5' region of the genome, where the open reading frames are located, than in the 3' UTR region. Surprisingly, Ouchterlony immunodiffusion assays failed to detect antigenic differences between these isolates. In general, the patterns of isolates from adjacent locations were more similar to each other than they were to isolates from more distant locations. A 1-km separation was sufficient for variants to be either similar to or highly variant from the type strain. At a single site isolates from adjacent plants were either identical or had related but variant genotypes. No more than two or three of the isolates were very similar to the type strain, suggesting a need for caution in accepting it as truly typical.

5 Experimental Manipulation of the STMV Genome

RNA transcripts from cDNA clones of the STMV genome are highly infectious (MIRKOV et al. 1990); this has facilitated analysis of genetic manipulations of the genome, which seems to display significant plasticity (ROUTH et al. 1995, 1997). Deletions and frameshift mutations in the first ORF (6.8-kDa protein) are readily tolerated, without any obvious effects on the ability of the mutants to replicate in protoplasts or whole plants, to accumulate virions, to spread systemically through the plant, and to influence symptom appearance. The function of this readily translated gene in the type strain remains unknown, but it is clearly dispensable for experimental infections.

Specific deletions and frameshift mutations in the coat protein, which should prevent the production of normal virions, are not lethal, but they do tend to compromise the virus such that accumulation of RNA signal is decreased in plants, but not so noticeably in protoplasts. Some of these mutants move systemically in plants and at a rate that resembles the movement of normal STMV. A by-product of the effort to design mutants needed to test the dispensability of the coat protein was the creation of variants that cause the induction of systemic veinal necrosis of tobacco, which can be very severe (ROUTH et al. 1995). This was an unexpected

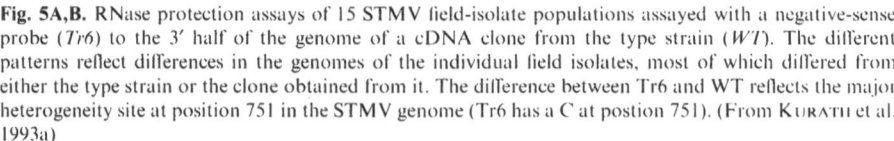

Fig. 5A,B. RNase protection assays of 15 STMV field-isolate populations assayed with a negative-sense probe (*Tr6*) to the 3' half of the genome of a cDNA clone from the type strain (*WT*). The different patterns reflect differences in the genomes of the individual field isolates, most of which differed from either the type strain or the clone obtained from it. The difference between Tr6 and WT reflects the major heterogeneity site at position 751 in the STMV genome (Tr6 has a C at postion 751). (From KURATH et al. 1993a)

A.

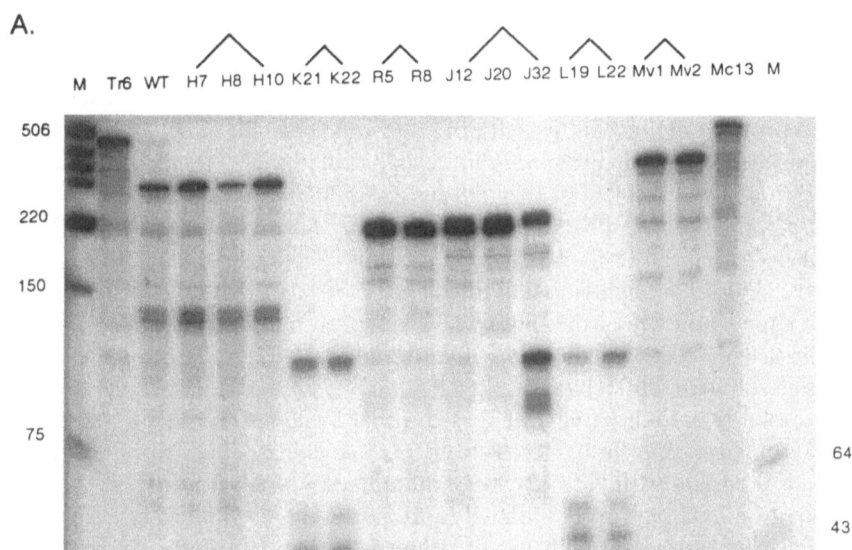

B.

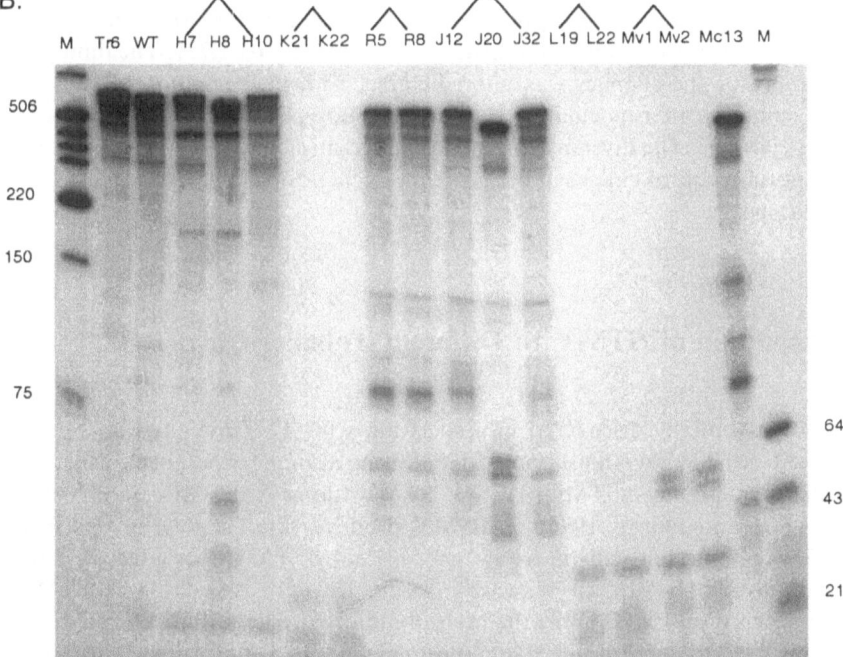

result, given the general absence of symptoms associated with natural and experimental infections of type STMV. The basis for these symptom effects of what are in effect unencapsidated RNA forms of STMV is unknown. A notable mutant in this series which had the above-described properties had 407 nucleotides deleted from the coat protein ORF, creating a variant that would code for only the first 13 native amino acids of the coat protein. Some viable mutants were made following manipulation of the 3' UTR, but they did not accumulate well, and the frequency of infection was low. It is possible to substitute the native 3' UTR of STMV with equivalent sequences from TMV, to form an infectious STMV/TMV chimera (KURATH et al. 1993), implying that the replicase of TMV is able to recognize either its own native sequence or the similar STMV sequence to initiate RNA replication.

The genetic changes that occur upon serial passage of characterized variants have been followed with STMV using RPA analysis of the entire genome (KURATH and DODDS 1995). Four parents were used, either uncloned genomic RNA from the type strain virus, infectious RNA from cDNA clones of a variant with base 751 = A, a variant with base 751 = C, or a mixture of RNA of the two base- 751 variants. In all, 42 lines were initiated (approximately ten siblings of each parent) and each was carried through ten or more passages in tobacco. Three different types of genetic change were observed, including the fixation of novel mutations in nine of 42 lines, mutation at the major heterogeneity site (base 751) in five of the 19 lines inoculated with a single genotype, and selection of a single major genome type in six of the 23 lines inoculated with mixed genotypes. Sequence analysis showed that the majority of mutations were single-base substitutions. The distribution of mutation sites included three clusters in which mutations occurred at or very near the same site, suggesting hot spots of genetic change in the STMV genome. One of these sites is near the major heterogeneity site at or near base 751. The timing of the appearance of the changes was unpredictable, with the exception of selection of single genotypes after inoculation with mixed genotypes, which usually happened at an early passage. The diversity of genetic changes in sibling lines is clear evidence of the important role of chance in the process of genetic diversification of STMV virus populations.

6 Adaptation of STMV to Different Tobamoviruses

STMV was initially discovered in association with TMGMV. It was quickly established that TMV would support STMV using a simple experiment, which was to inoculate an isolate of TMGMV and STMV along with TMV to *N. sylvestris* (VALVERDE and DODDS 1986). TMGMV produces local lesions in this host, so STMV would not normally move systemically either. TMV does infect *N. sylvestris* systemically, and STMV was detected in young leaves in association with TMV but in the absence of TMGMV in the multiply infected plants. Purified STMV RNA was later used to confirm this result and to extend the list of helper viruses beyond

TMGMV and TMV to include tomato mosaic virus, green tomato atypical mosaic virus, pepper mild mottle virus, and odontoglossum ringspot virus (VALVERDE et al. 1991; KURATH et al. 1992; RODRIGUEZ-ALVARDO et al. 1994). The legume strain of TMV, which is known to have a 3' UTR distinct from those of other tobamoviruses, did not support STMV.

Given these preliminary results, it was quite unexpected to find that several full-length cDNA clones that were highly infectious with TMGMV were not infectious with other tobamoviruses. Only one or two of ten plants in only one of several inoculations tested positive for STMV when TMV was used as the helper virus (KURATH et al. 1993a). These individual plants were retained, and the genome of the STMV that was able to replicate with TMV was characterized. In addition, STMV RNA from type strain virions, which is readily infectious with TMV, was passaged several times in the presence of TMV or other tobamoviruses. In both cases it was observed with RPA that a new variant of STMV had been selected or had adapted to TMV or other tobamoviruses. The genetic change mapped to a site in the 5' UTR, which was shown for variants adapted to TMV by sequence analysis to be the deletion of a single G in a series of 5 G residues at bases 61–65. Clones that captured this change were still not infectious with TMV, indicating that this deletion may be necessary but is not sufficient to permit STMV to replicate with TMV. Work in progress implicates the 5' terminal nucleotide, as well as the G residue at base 61 in the adaptation of STMV to TMV (Yassi and Dodds, unpublished). The infectious STMV/TMV chimera with a TMV 3' UTR was not supported by TMV but was supported by TMGMV.

Taken together, these results imply that the basis for adaptation of STMV to different tobamoviruses does not involve the ability of TMV replicase to make negative-sense RNA from genomic RNA and does involve the subsequent ability to make new positive-sense RNA. It is presumed that specific sequences or structures at the 5' end of the positive strand, or more likely the 3' end of the negative strand, are required for specific tobamovirus replicase recognition.

7 Conclusions

STMV is a well-characterized satellite virus that can be used as a model system for several types of virus studies. These include the role of RNA in the structure of spherical viruses, the development of foreign gene delivery based on viral genomes, genome plasticity and experimental virus evolution, natural genetic diversity of RNA viruses, and the general topic of subviral parasitism, including adaptation to alternate helper viruses.

156   J.A. Dodds

Acknowledgements. The author would like to acknowledge the research contributions made in his lab by J. Heick, R. Valverde, E. Mirkov, D. Mathews, G. Kurath, C. Rey, G. Rodriguez-Alvarado, G. Routh, and M. Ngon A Yassi. Additional collaborations with K. Kim (tissue ultrastructure), A. McPherson (particle structure), E. Mirkov (viral genetics), and A.L.N. Rao (protoplast studies) are greatly appreciated.

References

Dodds JA (1991) Structure and function of the genome of satellite tobacco mosaic virus. Can J Plant Pathol 13:192–195

Dodds JA (1993) DsRNA in diagnosis. In: Matthews REF (ed) Diagnosis of plant virus diseases. CRC Press, Boca Raton, pp 273–293

Felden B, Florentz C, McPherson A, Giege R (1994) A histidine accepting tRNA-like fold at the 3' end of satellite tobacco mosaic virus RNA. Nucleic Acids Res 22:2882–2886

Gultyaev A, van Batenburg E, Pleij C (1994) Similarities between the secondary structure of satellite tobacco mosaic virus and tobamovirus RNAs. J Gen Virol 75:2851–2856

Kim K, Valverde RA, Dodds JA (1989) Cytopathology of satellite tobacco mosaic virus and its helper virus in tobacco. J Ultrastruct Mol Res 102:196–204

Koszelak S, Dodds JA, McPherson A (1989) Preliminary analysis of crystals of satellite tobacco mosaic virus (STMV). J Mol Biol 209:323–326

Kurath G, Dodds JA (1994) Satellite tobacco mosaic virus sequence variants with only five nucleotide differences can interfere with each other in a cross protection-like phenomenon in plants. Virology 202:1065–1069

Kurath G, Dodds JA (1995) Mutation analyses of molecularly cloned satellite tobacco mosaic virus during serial passage in plants: evidence for hotspots of genetic change. RNA 1:491–500

Kurath G, Robaglia C (1995) Genetic variation and evolution of satellite viruses and satellite RNAs. In: Gibbs AJ, Calisher CH, Garcia-Arenal F (eds) Molecular basis of virus evolution. Cambridge University Press, Cambridge, pp 385–403

Kurath G, Rey CME, Dodds JA (1992) Analysis of genetic heterogeneity within the type strain of satellite tobacco mosaic virus reveals several variants and a strong bias for G to A substitution mutants. Virology 189:233–244

Kurath G, Heick JA, Dodds JA (1993a) RNAse protection analyses show high genetic diversity among field isolates of satellite tobacco mosaic virus. Virology 194:414–418

Kurath G, Rey CME, Dodds JA (1993b) Tobamovirus helper specificity of satellite tobacco mosaic virus involves a domain near the 5' end of the satellite genome. J Gen Virol 74:1233–1243

Larson SB, Koszelak S, Day J, Greenwood A, Dodds JA, McPherson A (1993a) Double-helical RNA in satellite tobacco mosaic virus. Nature 361:179–182

Larson SB, Koszelak S, Day J, Greenwood A, Dodds JA, McPherson A (1993b) Three dimensional structure of satellite tobacco mosaic virus at 2.9 A resolution. J Mol Biol 231:375–391

Mirkov TE, Mathews DM, Du Plessis DH, Dodds JA (1989) Nucleotide sequence and translation of satellite tobacco mosaic virus RNA. Virology 170:139–146

Mirkov TE, Kurath G, Mathews DM, Elliott K, Dodds JA, Fitzmaurice L (1990) Factors affecting efficient infection of tobacco with in vitro RNA transcripts from cloned cDNAs of satellite tobacco mosaic virus. Virology 179:395–402

Rodriguez-Alvarado G, Kurath G, Dodds JA (1994) Symptom modification by satellite tobacco mosaic virus in pepper types and cultivars infected with helper tobamoviruses. Phytopathology 84:617–621

Routh G, Dodds JA, Fitzmaurice L, Mirkov TE (1995) Characterization of deletion and frameshift mutants of satellite tobacco mosaic virus. Virology 212:121–127

Routh G, Ngon A Yassi M, Rao ALN, Mirkov TE, Dodds JA (1997) Cloned satellite tobacco mosaic virus specifically requires TMV-U5, but not its own intact coat protein, for replication in protoplasts of Nicotiana benthamiana. J Gen Virol 78:1277–1285

Valverde RA, Dodds JA (1986) Evidence for a satellite RNA associated naturally with the U5 strain and experimentally with the U1 strain of tobacco mosaic virus. J Gen Virol 67:1875–1884

Valverde RA, Dodds JA (1987) Some properties of isometric virus particles which contain the satellite RNA of tobacco mosaic virus. J Gen Virol 68:965–972

Valverde RA, Heick JA, Dodds JA (1991) Interactions between satellite tobacco mosaic virus, helper tobamoviruses and their hosts. Phytopathology 81:99–104

Luteovirus-associated Viruses and Subviral RNAs

B.W. Falk, T. Tian, and H.-H. Yeh

1 Introduction

Among the plant viruses, those in the family *Luteoviridae* (luteovirus is used here to refer to all definitive and tentative members of this family) are some of the most widespread and economically important (Miller 1994). They cause important diseases worldwide in nearly all of the crops human beings grow for food and fiber (Duffus 1977; Miller 1994). They are obligately vector transmitted from plant to plant by specific aphids, and, once acquired, a vector aphid may carry a corresponding luteovirus for the remainder of the aphid's life; the virus:vector transmission relationship is circulative:nonpropagative (Gildow 1987). Within the plant host the luteovirus infection is limited to the phloem tissues, and luteoviruses are generally classified as nonmechanically transmissible (D'Arcy et al. 1998).

Viruses in the family *Luteoviridae* are highly successful viruses, and many aspects of their molecular biology have been well studied in recent years (see Miller et al. 1997). Virions are isometric, approximately 30nm in diameter. Capsids are composed of two species of protein and all members of this family have relatively small ssRNA genomes ranging in size from 5.5 to 6.0kb (Miller et al. 1997; D'Arcy et al. 1998). Three distinct genome organizations are found for members of the family, and viruses with the respective genome organizations are

1 Shields Ave., Department of Plant Pathology, University of California, Davis, CA 95616, USA

divided into three genera: *Luteovirus, Polerovirus,* and *Enamovirus* (Fig. 1; D'ARCY et al. 1998). Viruses in the genera *Luteovirus* and *Polerovirus* have genomes which

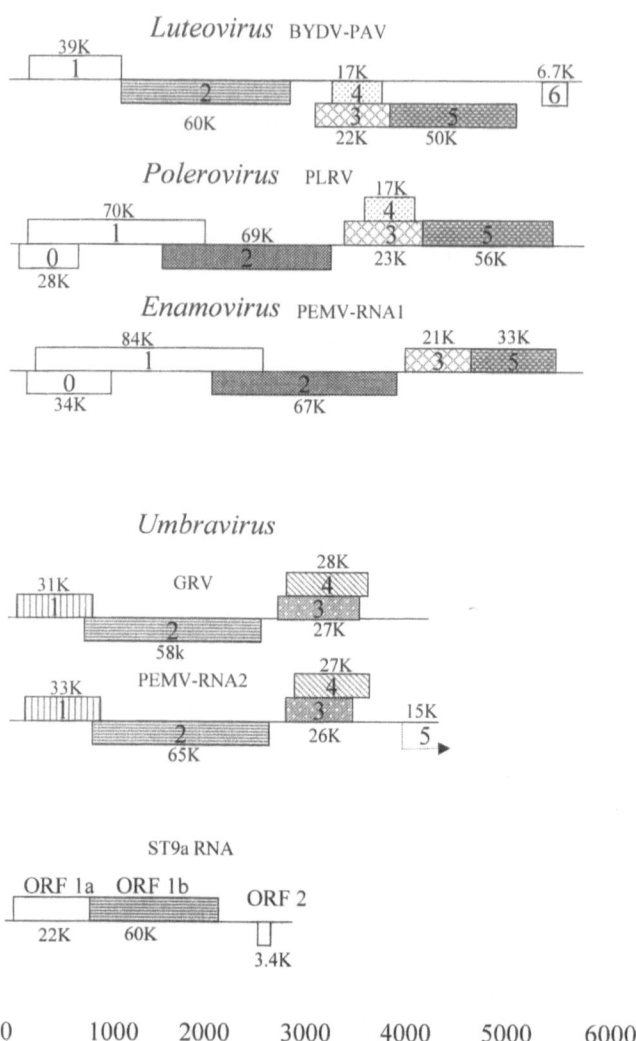

Fig. 1. Genome maps for luteoviruses, umbraviruses, and luteovirus-associated subviral RNAs. Boxes indicate ORFs, and similar shading indicates that the proteins encoded by respective ORFs have detectable similarity. Numbers in boxes indicate the given ORF, while numbers outside boxes indicate the sizes of encoded proteins (i.e., 39 K = 39 kDa). The upper three depict genome maps for genera of the family *Luteoviridae.* Generic names are followed by the corresponding species (e.g., BYDV-PAV is barley yellow dwarf virus PAV). The two middle maps are for groundnut rosette mosaic virus (GRV) and pea enation mosaic virus RNA2 (PEMV -RNA2), members of the genus *Umbravirus.* The bottom map represents the beet western yellows virus (BWYV) ST9-associated RNA (ST9aRNA). A scale (in nucleotides) is given at the bottom

encode all of the proteins required to initiate competent systemic plant infections, and which are required for aphid vector-mediated transmission from plant to plant. The single member of the genus *Enamovirus* appears to lack the genetic information to encode proteins facilitating systemic movement in planta (DEMLER et al. 1996) and is discussed below. The virion capsid protein(s) are the primary virus determinants which facilitate aphid-transmission specificity of luteoviruses (ROCHOW 1970; 1977; GILDOW 1987; DEMLER et al. 1997). Indeed, the luteoviruses are worthy of study because of their importance as plant pathogens and because of the complex interactions they undergo with their host plants as well as with their aphid vectors. However, it is because of interactions which occur in mixed infections between luteoviruses and other viruses and between luteoviruses and subviral RNAs that they are discussed here.

Luteoviruses commonly occur in mixed infections (ROCHOW 1977; FALK and DUFFUS 1981; CREAMER and FALK 1990). The mixed infections occur in crop and weed host plants, between related and unrelated luteoviruses (CREAMER and FALK 1990), and, as will be discussed further here, between luteoviruses and unrelated viruses or between luteoviruses and subviral RNAs. In many cases, these latter mixed infections are specific associations (virus complexes) that often result in distinct diseases which differ in symptom type and severity from diseases caused by any of the viruses when they infect plants alone. Different types of viruses and/or subviral RNAs can be associated with luteoviruses in these complexes. Despite the different characteristics and molecular or taxonomic properties of these luteovirus-associated viruses and subviral RNAs, they all take advantage of the co-infecting luteovirus as a means of gaining their own aphid transmissibility and thereby dispersal to new host plants (therefore survival!). Since the discovery of these virus complexes, their biological characteristics have been studied in some detail. We will concentrate here on recent information learned about the molecular biology of the specific viruses and/or subviral RNAs in the luteovirus-associated complexes and their interactions with their associated luteoviruses.

2 Helper-dependent Aphid-transmitted Virus Complexes

The first luteovirus-associated aphid-transmitted virus complex was recognized in 1946. K.M. SMITH (1946) reported that two distinct viruses, when co-infecting tobacco plants, caused the tobacco rosette disease. Aphid transmission experiments using *Myzus persicae* from tobacco rosette-affected plants generally gave typical tobacco rosette symptoms on inoculated tobacco plants. However, aphid transmission experiments occasionally yielded plants which did not show rosette but rather vein distortion symptoms. Smith named this virus tobacco vein distorting virus (TVDV). TVDV could be continuously transmitted by *M. persicae* and always gave only the vein distortion symptom type on tobacco plants. Its biological properties are typical of viruses in the genus *Polerovirus*, although as yet it is not a formally recognized member of that genus. Smith also noted that mechanical transmission experiments from rosette-affected tobacco plants resulted in tobacco plants exhibiting milder symptoms than those seen for tobacco rosette and different from those seen for infections by only TVDV. Plants exhibited mild but distinct mottling. Surprisingly, *M. persicae* was unable to transmit the virus from tobacco plants showing only the mild mottle. Smith named the mechanically transmissible virus tobacco mottle virus (TMoV); now a member of the genus *Umbravirus*. He concluded that tobacco rosette is caused by co-infections of TMoV and TVDV, and that TMoV and TVDV can both be transmitted by *M. persicae* from doubly infected tobacco plants.

2.1 Umbraviruses are Helper Dependent

Since the discovery of tobacco rosette by Smith, several other luteovirus-associated aphid-transmitted virus complexes have been described, and many cause important diseases (see Table 1). One of the most widespread and best studied is carrot motley dwarf (STUBBS 1948), shown to be caused by co-infection with carrot redleaf virus (CRLV, an ungrouped member of the *Luteoviridae*) and the *Umbravirus* carrot mottle virus (CMoV; WATSON et al. 1964). Carrot motley dwarf and most of the virus disease complexes listed in Table 1 show properties very similar to those described above for tobacco rosette; they are comprised of two viruses: one typical

Table 1. Luteovirus: *Umbravirus* and luteovirus:subviral RNA virus complexes

Viruses or subviral RNA[a]	Host plant	Effects[b]
Luteovirus:*Umbravirus*		
BWYV + LSMV	Dicots	Aphid transmissibility
CRLV + CMoV	Dicots	Aphid transmissibility
GRAV + GRV	Dicots	Aphid transmissibility
TVDV + TMoV	Dicots	Aphid transmissibility
TVAV + TYVV	Dicots	Aphid transmissibility
PEMV RNA1 + PEMV RNA2	Dicots	Aphid transmissibility
		Systemic movement
Luteovirus:subviral RNA		
BWYV + ST9aRNA	Dicots	Aphid transmissibility
		Systemic movement
Luteovirus:*Umbravirus*:subviral RNA		
CRLV + CMoV +	Dicots	Aphid transmissibility
CRLVaRNA		Systemic movement

[a]The viruses and subviral RNAs comprising the specific complexes are indicated. For the luteovirus:*Umbravirus* complexes, the first acronym indicates the luteovirus and the second indicates the *Umbravirus*. For example BWYV and LSMV are the *Polerovirus* beet western yellows luteovirus and the *Umbravirus* lettuce speckles mottle umbravirus, respectively. Other acronyms are defined in the text. For luteovirus: subviral RNA complexes, ST9aRNA is the BWYV ST9-associated RNA. For luteovirus:*Umbravirus*:subviral RNA complexes, CRLVaRNA is the carrot redleaf virus associated RNA.
[b]Aphid transmissibility denotes that the second virus or subviral RNA acquires the trait only as a result of being in the mixed infection with the first. Systemic movement indicates that the second virus or subviral RNA does not readily move systemically in the given host unless co-infecting with the virus listed first. However, for PEMV the second virus confers cell-to-cell and movement ability to the first.

luteovirus and one *Umbravirus*. Because the *Umbravirus* member of the complex is not aphid transmissible from single infections but is dependent on being in a mixed infection with a luteovirus, the luteovirus is referred to as the helper virus (helper for aphid transmission!). The *Umbravirus* is often referred to as the dependent virus, being dependent on the luteovirus only for aphid transmission. The *Umbravirus* is competent for replication, and for cell-to-cell and systemic movement within the host plant. Together the two viruses compose the helper-dependent aphid-transmitted virus complex.

It is now well established that the mechanism of helper-dependent aphid transmission of the *Umbravirus* from doubly infected plants is a result of trans-capsidation or genomic masking (Rochow 1977; Falk et al. 1979; Waterhouse and Murant 1983). During simultaneous replication of the luteovirus and the *Umbravirus* within cells of the doubly infected plant, structural interactions result during virion assembly and progeny virions are formed. Some of the progeny *Umbravirus* genomic RNA becomes encapsidated by capsid proteins encoded by the co-infecting luteovirus (Fig. 2). Thus, there are two types of virions within the doubly infected cell: those of the progeny luteovirus, and those which are composed of luteovirus capsids but contain the *Umbravirus* genomic RNA. As the luteovirus virion capsid proteins are virus-specific aphid-transmission determinants (Rochow 1977; Gildow 1987), when the *Umbravirus* genomic RNAs become encapsidated by luteovirus capsid proteins, these genomically masked RNAs gain the aphid-transmission characteristics of the corresponding co-infecting helper luteovirus. Following aphid acquisition, both viruses are most often simultaneously transmitted by aphids to subsequent host plants. It is highly likely that there must be specific signals or sequences which facilitate RNA encapsidation by the luteovirus capsid proteins. However, encapsidation signals have not yet been identified, and at least in one case more than just simultaneous co-replication of a luteovirus and *Umbravirus* is required. Murant (1990) has shown that aphid transmissions of the *Umbravirus* groundnut rosette virus (GRV) result only from plants triply infected by GRV, the helper virus groundnut rosette assistor virus (GRAV), and a GRV satellite RNA. The role of the GRV satellite RNA in assisting genomic masking and/or subsequent aphid transmissibility of GRV is not known.

2.2 *Umbravirus* Helper-dependent Specificity

The luteovirus-*Umbravirus* interactions which occur in a specific disease complex (i.e., tobacco rosette) are not limited to the original co-infecting viruses. This has been shown when specific umbraviruses have been inoculated to plants along with other luteoviruses (Adams and Hull 1972; Falk et al. 1979; Waterhouse and Murant 1983). Adams and Hull (1972) used the viruses from the groundnut rosette and tobacco yellow vein complexes and created heterologous mixed infections of GRAV plus the *Umbravirus* tobacco yellow vein virus (TYVV). When they performed aphid transmissions from these plants by using the GRAV aphid vector, *Aphis craccivora*, they found that GRAV was able to serve as a helper virus for the

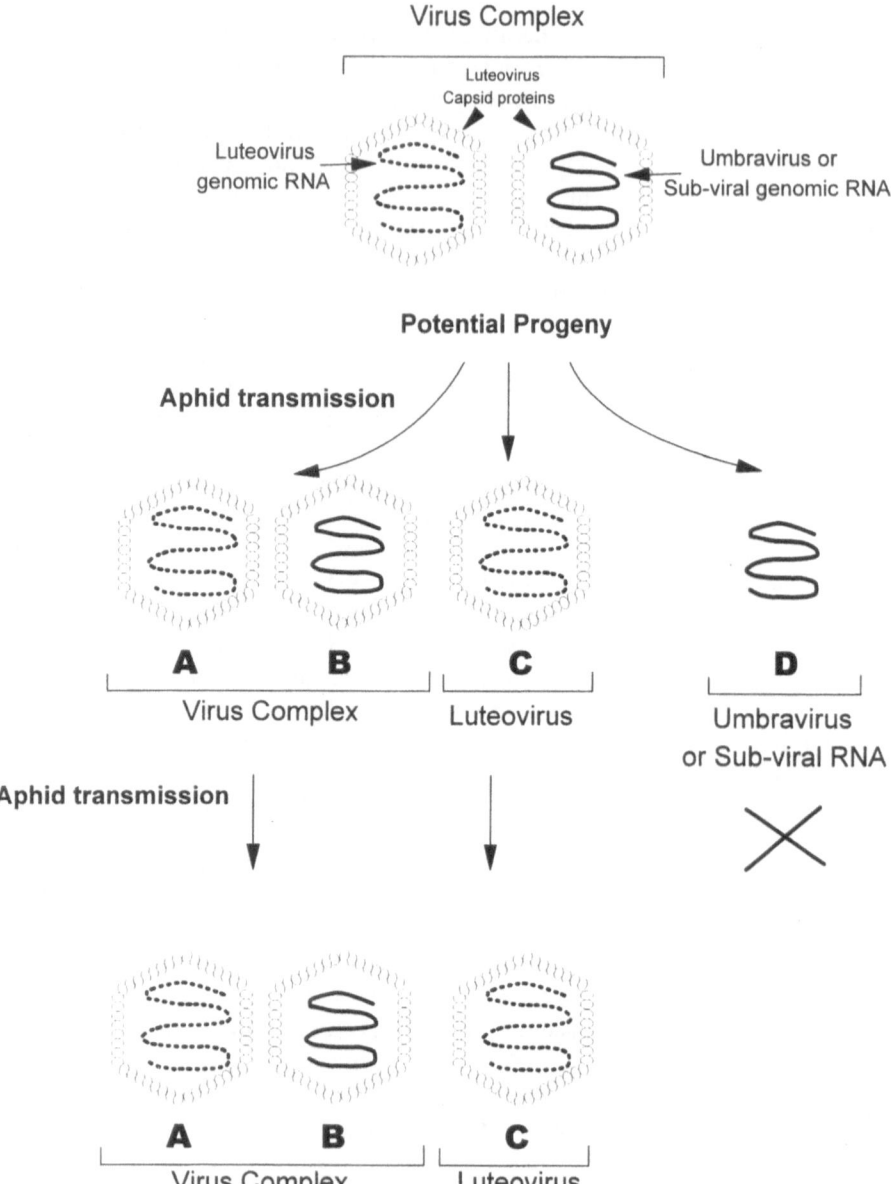

Fig. 2. Transcapsidation interactions for the luteovirus helper-dependent virus/subviral RNA complexes. The original virus complex is depicted at the *top*. Both types of virus particles contain capsids composed of luteovirus capsid protein molecules. Aphid transmission to new plants could result in the potential progeny shown: the original virus complex; only the luteovirus member; or infections of only the umbravirus or the subviral RNA. The virus complex and the luteovirus would yield competent infections which could be further transmitted by aphids. Infections of only the umbravirus could be systemic infections, but as no luteovirus helper is present, these would be dead ends. Infections by only the subviral RNA are likely to be limited to only one or a few cells and represent dead ends

transmission of TYVV. TYVV is normally associated with tobacco yellow vein assistor virus (TVAV) in the tobacco yellow vein complex and, together, both viruses are transmitted by *M. persicae*. However, when TYVV co-infects plants with the GRAV, both are transmissible from the co-infected plant by the GRAV aphid vector, *A. craccivora*. Similar experiments have been performed by WATERHOUSE and MURANT (1983) using the *Umbravirus*, CMoV (normally transmitted by *Cavariella aegopodii* when co-infecting plants with CRLV) and potato leafroll virus (PLRV), the type member of the genus *Polerovirus*.

This ability to utilize more than one helper luteovirus (and then more than one aphid vector) is likely a significant epidemiological advantage for viruses in the genus *Umbravirus*. The host ranges of the helper luteovirus and *Umbravirus* for a given complex are not identical (ADAMS and HULL 1972; FALK et al. 1979), yet common hosts for both viruses are necessary for success of the virus complex. For example, if the viruliferous vector aphid carrying both viruses of the complex feeds on a plant which is a host for only one of the viruses, different scenarios may result. If the inoculated plant is a host for the luteovirus but not the *Umbravirus*, the luteovirus will infect the plant and it may subsequently be transmitted by aphids to new host plants, but the *Umbravirus* will be lost. However, if the converse were to occur, the *Umbravirus* could initiate an infection but the luteovirus would be lost. The *Umbravirus* could replicate and systemically spread throughout the host plant, but no virions would be formed and subsequent aphid transmission would not occur. In a sense, this would represent a dead end host for the *Umbravirus*. However, if another luteovirus were then to infect the *Umbravirus*-infected plant, this could result again in a competent interaction, giving rise to encapsidation of the *Umbravirus* genomic RNA by the capsid proteins of the co-infecting luteovirus. This could result in the genomically masked *Umbravirus* RNA being aphid transmissible, perhaps even by a new aphid vector, that of the co-infecting luteovirus.

2.3 *Umbravirus* Genome Organization

The genomes of three viruses in the genus *Umbravirus* [groundnut rosette virus (GRV), carrot mottle virus (CMoV) and pea enation mosaic virus (PEMV) RNA2] have been sequenced and their organizations described (DEMLER et al. 1993; TALIANSKY et al. 1996; GIBBS et al. 1996). The ssRNA genomes of CMoV and GRV are small, only 4201 and 4019 nucleotides, respectively. Their organizations are quite similar, each containing four ORFs (see Fig. 1). The genome organization of PEMV RNA2 is slightly different and will be discussed below. By comparing deduced amino acid sequences of the putative proteins encoded by the CMoV and GRV genomes with those in data bases, tentative functions were identified for some proteins. Comparisons showed that the CMoV and GRV ORF 1-encoded proteins were similar to each other and to the ORF 1 protein of PEMV RNA2, but no significant similarities with other proteins in the data bases were identified. However, the ORF 2-encoded protein showed a high degree of similarity with the RNA-dependent RNA polymerases of viruses in the genera *Dianthovirus*, *Carmovirus*,

Necrovirus, and *Luteovirus* (TALIANSKY et al. 1996; GIBBS et al. 1996). ORFs 1 and 2 are overlapping and the overlapping sequence contains the consensus heptanucleotide (AAAUUUU for GRV) which is characteristic of −1 ribosomal frameshifts (MILLER et al. 1997). Based on the similarities to other plant viruses and the presence of the shifty heptanucleotide, these data suggest that during translation of the genomic RNA, a −1 ribosomal frameshift likely results, to yield the RNA-dependent RNA polymerase. Thus, both ORFs 1 and 2 encode the RNA-dependent RNA polymerase. ORFs 3 and 4 also are overlapping and those of GRV encode for proteins of approximately 27 kDa and 28 kDa, respectively (TALIANSKY et al. 1996). Initial comparative sequence analyses failed to suggest a possible function for the ORF 3 putative protein but led to suggestions that the CMoV and GRV ORF 4 putative proteins were probably involved in cell-to-cell movement, as a high degree of similarity was noted for these and the 3a proteins of cucumoviruses and bromoviruses (TALIANSKY et al. 1996; GIBBS et al. 1996). Recently, additional supporting evidence was obtained by expressing the GRV ORF 4 protein in plants by using the potato virus X vector (RYABOV et al. 1998). Chimeric proteins composed of GFP (the green fluorescent protein) and the GRV ORF4 protein accumulated in plant cells at cell walls near plasmodesmata, suggestive of a possible role in cell-to-cell movement. More definitive evidence supporting the hypothesis that the GRV ORF 4 protein directs cell-to-cell movement was obtained by constructing PVX mutants lacking functional coat protein (CP) and triple gene block regions. These PVX mutants cannot move from cell to cell in infected plants. However, when these mutants were engineered to also express the GRV ORF 4 protein, the ability to move from cell to cell was restored, providing further evidence that the GRV ORF 4 is a cell-to-cell movement determinant (RYABOV et al. 1998).

Viruses in the genus *Umbravirus* can replicate and move from cell to cell and systemically within plants. These properties are not supplied by the helper virus, but proteins. determining these functions are encoded within the *Umbravirus* genome. However, they do not encode a capsid protein. Thus, as suggested by the biological data, their sole means of encapsidating their genomic RNAs is to co-infect plants with their helper luteovirus. The *Umbravirus* genomic RNA is then encapsidated via genomic masking by using the capsid proteins supplied by the luteovirus, but, probably of more importance, they also gain the ability to be dispersed via aphid transmission to new plant hosts.

2.4 Pea Enation Mosaic Virus

Pea enation mosaic is a complex virus disease which in many ways resembles the helper-dependent aphid-transmitted virus complexes described above. It is caused by co-infecting viruses, or viral RNAs, but, in contrast to the *Umbravirus*:luteovirus associations, the co-infecting pea enation mosaic virus (PEMV) RNAs appear to have a definitive mutual dependence on each other (see review by DEMLER et al. 1996). The two viruses/RNAs are PEMV RNA 1, the sole member of the genus *Enamovirus* within the family *Luteoviridae*, and PEMV RNA 2, a member of the genus *Umbravirus*. The PEMV RNA 1 genome is an ssRNA of 5705 nucleotides and its organization resembles the genomic organization of viruses in the genus *Polerovirus* (Fig. 1). PEMV RNA1 contains ORFs encoding proteins which facilitate its replication (ORFs 1 and 2) and encapsidation (ORFs 3 and 5), but it apparently does not encode functional proteins that can facilitate cell-to-cell or systemic spread. PEMV RNA 1 is competent for replication within protoplasts but fails to systemically infect plants when inoculated alone (DEMLER et al. 1994). PEMV RNA 1 is dependent for systemic spread in plants by co-infecting them with PEMV RNA 2. Interestingly, PEMV RNA 1 lacks an ORF homologous to the ORF 4 of viruses in the genus *Polerovirus*. Recent data suggest that for the *Polerovirus* PLRV, the protein encoded by ORF 4 is one of the proteins involved in systemic movement (SCHMITZ et al. 1997).

PEMV RNA 2 is an ssRNA of 4253 nucleotides (DEMLER et al. 1993). Like PEMV RNA 1, PEMV RNA 2 alone is competent for replication. However, PEMV RNA 2 also has the ability to systemically infect plants in the absence of RNA 1, but no virions are produced in the PEMV RNA 2 infections. The co-infection of plants by both PEMV RNAs leads to systemic infection by both and yields virions separately containing either PEMV RNA1 or PEMV RNA2 (DEMLER et al. 1996).

Similar to other viruses in the genus *Umbravirus*, the PEMV RNA 2 sequence contains four complete ORFs. However, in contrast to the organizations seen for GRV and CMoV, a fifth potential ORF which continues to the 3' terminus of PEMV RNA 2 is also present (Fig. 1). ORFs 1 and 2 likely encode an RNA-dependent RNA polymerase, as sequence comparisons of the deduced proteins have identified some of the characteristic features seen for RNA-dependent RNA polymerases. The putative proteins encoded by these ORFs share similarity to the respective proteins encoded by GRV and CMoV (TALIANSKY et al. 1996; GIBBS et al. 1996). Also as for GRV and CMoV, the function of the ORF 3-encoded protein is as yet unknown, but computer-assisted sequence comparisons suggest that ORF 4 encodes a protein involved in cell-to-cell movement (DEMLER et al. 1996). Analysis of the PEMV RNA 2 sequence shows that PEMV RNA 2 does not encode a capsid protein. The encapsidation by the PEMV RNA 1-encoded capsid protein(s) yields aphid-transmissible PEMV RNAs 1 and 2. In turn, PEMV RNA2 encodes for protein(s) which facilitate cell-to-cell (and systemic?) movement of both PEMV RNAs. Whether the PEMV complex can be called bipartite or a mixed infection is somewhat debatable (DEMLER et al. 1996). Even though each PEMV RNA encodes proteins which facilitate their own replication, it is clear that both RNAs are necessary for the survival of the complex.

3 Luteovirus-associated Subviral RNAs

More recently, another type of luteovirus:helper-dependent virus/RNA association has been identified. The first of these to be described was for a member of the genus *Polerovirus*, beet western yellows virus (BWYV) ST9 (FALK and DUFFUS 1984). This 'strain' of BWYV was identified because of the unusual symptoms induced by BWYV ST9 infection in the common BWYV indicator plant, *Capsella bursa-pastoris* (shepherd's purse). Shepherd's purse plants infected by BWYV ST9 are severely stunted compared with the more typical overall yellowing induced by infection with common isolates of BWYV (FALK and DUFFUS 1984; SANGER et al. 1994). The biologies of BWYV ST9 and common BWYV, however, were generally similar. No mechanical transmission was ever obtained from BWYV ST9 or common BWYV-infected plants. Nearly identical aphid-transmission characteristics were also seen, with the exception that aphid transmissions of BWYV ST9 to shepherd's purse plants consistently gave a few plants (3% or less) which showed typical BWYV symptoms, not the severe stunting typical of BWYV ST9. Subsequent aphid transmissions from these plants never again resulted in severe symptoms typical of BWYV ST9.

Analysis of the virus-specific double-stranded RNAs from BWYV-ST9-infected plants and of virion BWYV ST9 RNAs showed that BWYV ST9 plants contained a relatively abundant ~2.8-kb RNA which was not present in virions of, or plants infected by, common BWYV (FALK and DUFFUS 1984). The 2.8-kb ssRNA was encapsidated in virions composed of BWYV capsid proteins, and nucleic acid hybridization experiments showed that it lacked sequence homology with the BWYV genomic RNA (FALK et al. 1989). We named the 2.8-kb RNA the ST9-associated RNA (ST9aRNA).

Nucleotide sequence analysis of cloned cDNAs revealed the ST9aRNA to be an ssRNA of 2844 nucleotides (CHIN et al. 1993; FALK and TIAN 1998; WATSON et al. 1997). Only two ORFs are contained within the sequence, and the RNA contained within virions is of positive polarity. The first ORF can encode a protein of 22,134 mW. However, this ORF contains an amber stop codon (UAG) at nucleotide 629, and immediately beyond the UAG the ORF continues to a UAA stop codon at position 2210. The nucleotide sequence surrounding the UAG codon exhibits similarities with translational readthrough codons found for many plant viruses, suggesting that it may be a functional readthrough codon (CHIN et al. 1993). Furthermore, in vitro translation analyses showed that proteins of ~22kDa and 82kDa were generated when the ST9aRNA was translated using wheat germ or rabbit reticulocyte lysate extracts (CHIN et al. 1993). The 82-kDa protein most likely results from translational readthrough. Thus ORF 1 can be divided into two parts, ORF 1a (22kDa) and ORF 1b (60kDa), and via translational readthrough they encode the 82kDa protein. Computer-assisted analyses of the putative proteins encoded by ORFs 1a and 1b showed that the ORF 1a-deduced amino acid sequence had no significant similarity to other proteins in existing data bases. However, the ORF 1b-deduced amino acid sequence exhibits a high degree of similarity with RNA-dependent RNA polymerases, particularly those of viruses in the family *Tombusviridae*, and to the *Luteovirus* BYDV-PAV (CHIN et al. 1993).

ORF 2 is located downstream of ORF 1 and potentially encodes a protein of only 3386 mW (3.4kDa; CHIN et al. 1993). It is not yet known if this small ORF is functional. However, current evidence suggests that it, or this region of the ST9aRNA, may be important. First, a highly abundant subgenomic RNA which maps to this region is generated in plants and protoplasts infected by the ST9aRNA (FALK et al. 1989; CHIN et al. 1993; PASSMORE et al. 1993). Second, as described below, a newly discovered RNA that is highly similar to the ST9aRNA also has a small ORF located at roughly the same position (WATSON et al. 1997). Nucleotide sequence analyses also show that the ST9aRNA does not encode any proteins with similarities to the capsid proteins or cell-to-cell movement proteins of other plant viruses. Thus, like the viruses in the genus *Umbravirus*, the ST9aRNA most likely relies on co-infections with its BWYV helper virus to supply capsid proteins for encapsidation of the ST9aRNA. However, unlike viruses in the genus *Umbravirus*, the ST9aRNA appears to be incapable of independent cell-to-cell or systemic movement within plants. Thus, it appears to also depend on the BWYV helper virus for this function, although the mechanism is as yet unknown.

To determine whether or not the ST9aRNA was capable of independent replication, we used full-length transcripts derived from cloned ST9aRNA cDNAs to inoculate *Nicotiana tabacum* protoplasts in the presence and absence of infectious BWYV full-length transcripts. These experiments showed that the ST9aRNA is competent for replication in the absence of the helper virus BWYV (PASSMORE et al. 1993). The ST9aRNA replicated to similar levels whether or not the BWYV was present. Interestingly, our analyses showed that the ST9aRNA stimulated the accumulation of the helper virus (BWYV) RNA in doubly infected protoplasts (PASSMORE et al. 1993). This also supports results from whole plant experiments, which showed that the titers of the BWYV genomic RNA and capsid proteins are approximately ten times higher in BWYV-ST9-infected plants than they are in plants infected by common BWYV (FALK and DUFFUS 1984; SANGER et al. 1994).

Having full-length cDNA clones for both BWYV and the ST9aRNA also allowed us to examine more closely factors affecting the increased accumulation of BWYV in plants. *N. tabacum* protoplasts were inoculated using combinations of transcripts and the ST9aRNA transcripts plus common BWYV virion RNA. Progeny virions were then harvested from protoplasts and fed to *M. persicae* aphids using the in vitro parafilm membrane system (FALK et al. 1979; SANGER et al. 1994). Whenever the ST9aRNA was present with BWYV (either BWYV infectious transcripts or the common BWYV genomic RNA), the resulting shepherd's purse plants showed the severe stunting symptoms typical of the wild-type BWYV ST9 infection, and the BWYV titer was approximately tenfold that in plants infected only by BWYV (SANGER et al. 1994). Transmission electron microscopy and immunogold labeling analyses showed that these infections were limited to the vascular tissues, whether or not the ST9aRNA was present. Thus, these data suggest that co-infection of plant cells by BWYV and the ST9aRNA results in a significant (approximately tenfold) increase in BWYV genomic RNA and capsid protein antigens per cell, and there is no escape from the vascular tissues in doubly infected plants. The severe symptom phenotype seen in BWYV ST9-infected shepherd's purse plants may be a result of the increased accumulation of BWYV per cell.

The ST9aRNA was further analyzed by constructing specific mutants and comparing their replication in protoplasts. Frameshift and specific point mutations were introduced into the ORF 1a, ORF 1b, and ORF 2 regions. Point mutations were introduced either to eliminate start codons or to introduce silent marker (changing nucleotides and not the resulting amino acids) mutations into the ST9aRNA sequence. Our initial results indicate that frameshift mutations which disrupted ORFs 1a and 1b yielded RNAs that were noninfectious. However, silent marker mutations for these same ORFs resulted in mutants that replicated essentially equal to (ORF 1a) or only slightly less than (ORF 1b) wild type. When we eliminated the ORF 2 AUG, this mutant was still replication competent, but to very low levels compared with wild type.

Recent evidence suggests that luteovirus-associated subviral RNAs such as the ST9aRNA are more widespread than is presently recognized. An RNA very similar to the ST9aRNA was recently discovered as part of the California carrot motley dwarf virus complex and has been named the carrot redleaf virus (CRLV; a member of the *Luteoviridae*)-associated RNA (CRLVaRNA; WATSON et al. 1997). The biological and molecular characteristics of the CRLVaRNA very closely resemble those of the ST9aRNA, except that in natural infections it is associated with CRLV and the *Umbravirus* CMoV (WATSON et al. 1997). Despite repeated attempts, we were unable to demonstrate that co-infection with only CMoV was sufficient to allow systemic infection of plants by CRLVaRNA, suggesting that the CRLVaRNA cannot utilize the *Umbravirus* CMoV as a helper virus for systemic infection. Like the ST9aRNA, the CRLVaRNA is dependent on a helper virus, CRLV, both for systemic spread within the co-infected plant and for aphid transmission from plant to plant. The mechanism of the latter is heterologous encapsidation, but the mechanism(s) involved in systemic spread within the plant host is as yet unknown.

We used cloned cDNAs and northern hybridization analysis to assess the natural incidences in crop and weed host plants of the ST9aRNA and CRLVaRNA in California. Initial sampling showed that 20 of 126 cruciferous weed samples were found to be infected by BWYV + the ST9aRNA, while 28 of these 126 plants were infected by only BWYV (FOSTER 1989). Thus, 38% of these plants were infected by BWYV, but 42% of the BWYV infections also contained the ST9aRNA. Furthermore, in our studies on California carrot motley dwarf the CRLVaRNA was always associated with CRLV and CMoV in carrot motley dwarf-affected carrots, including those analyzed from three distinct geographic regions in California (WATSON et al. 1997). The CRLVaRNA was also recently identified from infected parsley in Belgium (Vercruysse, personal communication), showing that it may be widespread geographically. Thus, these and other luteovirus-associated subviral RNAs may be more common than is currently recognized.

Like viruses in the genus *Umbravirus*, the ST9aRNA and CRLVaRNA are not restricted in their ability to use other viruses as helper viruses. We created mixed infections of BWYV ST9 and the California carrot motley dwarf viruses in chervil (*Anthriscus cerefolium*) plants. Then, *M. persicae* and *C. aegopodii* aphids were separately used to recover and transmit viruses from the mixed infections. Northern hybridization analysis using specific probes for each of the five possible RNAs showed that the subsequently inoculated plants contained various combinations of the original viruses and subviral RNAs. These experiments clearly demonstrated that the CRLVaRNA can utilize BWYV as a helper virus, and that when it does, it then is transmissible by the BWYV aphid vector, *M. persicae*. Similarly, the ST9aRNA was able to utilize CRLV as a helper virus, and in these cases it was then transmissible by the CRLV aphid vector, *C. aegopodii* (Falk, unpublished).

4 Conclusions and Final Thoughts

Viruses in at least two genera (*Polerovirus* and *Enamovirus*) of the family *Luteoviridae* appear to be quite prolific in their ability to acquire and support helper-dependent viruses and/or subviral RNAs. It is interesting that the types of helper-dependent luteovirus-associated viruses and/or subviral RNAs can be quite different, representing different virus taxa. The luteovirus-associated subviral RNAs discussed here are generally distinct from other types of helper-dependent subviral RNAs such as satellite viruses and/or RNAs, in that they do not depend on the helper virus for replication (exceptions are the satellite RNA associated with the *Polerovirus* BYDV-RPV [MILLER et al. 1991; RASOCHOVA et al. 1997] and the satellite RNAs associated with Umbraviruses, and thus indirectly with luteoviruses [DEMLER and DE ZOETEN 1989; MURANT 1990]). Of the viruses and subviral RNAs discussed here, some have the ability to establish systemic infections in plants in the absence of their helper luteoviruses (i.e., those in the genus *Umbravirus*) and some do not (the subviral RNAs), but all are dependent on capsid proteins, which they gain from the co-infecting helper virus via transcapsidation. As a result, they also gain an essential biological property: the ability to be disseminated to new plants by the aphid vectors of the co-infecting helper virus. In turn, the obvious question is: "Do the helper-dependent viruses or subviral RNAs provide any benefit to their co-infecting helper viruses?"

The association of PEMV RNA 1 and PEMV RNA 2 to comprise the PEMV complex is the most clear-cut example of mutual benefit. PEMV RNA 2 gains aphid transmissibility and dispersal via transcapsidation. However, PEMV RNA 1 gains the ability to move from cell to cell in nonphloem tissues and systemically through the plant as a result of its association with PEMV RNA 2. The advantage here is that the luteovirus (*Enamovirus*) PEMV RNA 1 not only gains the ability to move systemically through phloem tissues, but also moves into and through mesophyll tissues. As a result, this has loosened restrictions on the ability of PEMV RNAs 1 and 2 to be transmitted by aphids to new host plants. The viruses in the family *Luteoviridae* are transmitted from plant to plant by phloem-feeding aphids in a circulative-nonpropagative manner (MILLER 1994). The PEMV RNAs are acquired from infected plants by feeding *Acyrthosiphon pisum* aphids and are retained essentially for life – typical properties of circulative-nonpropagative aphid transmission (NAULT and GYRISCO 1966). Despite being acquired by feeding from the phloem, PEMV RNAs 1 and 2 are not restricted to the phloem for aphid inoculation. PEMV RNAs 1 and 2 can be transmitted to plants by viruliferous aphids in much shorter times than are required for aphids to reach the phloem tissues. NAULT and GYRISCO (1966) and TOROS et al. (1978) reported that short inoculation access periods were sufficient for aphid transmission of PEMV. NAULT and GYRISCO (1966) showed that inoculations could take place in as little as 7s, during intercellular aphid sampling probes in the epidermal tissues, and a 7.2% transmission rate was observed for probes ranging between 6 and 12 s. They also showed that *A. pisum* required at least 5 min to reach phloem tissues. Thus, in these short inoculation access periods aphids were transmitting the PEMV RNAs to epidermal and/or mesophyll tissues, and because PEMV RNA 2 encodes a protein (the ORF 4 protein) which facilitates the cell-to-cell movement of both RNAs, the infection readily spread throughout the plant, including into phloem tissues. Thus, even though the acquisition and retention phases are typical of circulative-nonpropagative aphid transmission, the inoculation phase is not. The inoculation threshold is much shorter, almost resembling that of nonpersistent aphid transmission.

In the other luteovirus:*Umbravirus* complexes such as carrot motley dwarf it also is advantageous (essential) for the *Umbravirus* to be associated with the luteovirus. However, the advantage for the co-infecting luteovirus is not so obvious. Unlike the PEMV complex, in general the luteovirus members (i.e., CRLV) of complexes such as the carrot motley dwarf complex do not gain the ability to move cell to cell in mesophyll tissues. Even though the *Umbravirus* members can move efficiently cell to cell in mesophyll tissues, the luteoviruses apparently cannot efficiently take advantage of the *Umbravirus* cell-to-cell movement protein. Thus, the luteoviruses appear to still be phloem-limited in these co-infections. However, FALK et al. (1979) reported limited mechanical transmission of the *Polerovirus* BWYV only when it was associated in co-infections with the *Umbravirus* lettuce speckles mottle virus (LSMV). This suggests that BWYV mesophyll cell-to-cell movement may occasionally result from co-infections of BWYV and LSMV, presumably by LSMV complementing BWYV cell-to-cell movement. This was not always the case. In addition, BARKER (1989) reported that co-infections of the *Umbravirus* CMoV and the *Polerovirus* PLRV in *Nicotiana clevelandii* plants resulted in an inconsistent but reproducible increase in PLRV antigen titer by as much as tenfold. This was not a specific effect, in that several other mechanically transmissible plant viruses were able to induce the same effects on PLRV titer in co-infections. Whether or not Umbraviruses stimulate titer of their respective helper viruses in co-infected plants, and how this is accomplished, are important questions remaining to be answered.

The ST9aRNA and CRLVaRNA are subviral coat-dependent RNA replicons (PASSMORE et al. 1993) and as yet are not assigned to any formal taxonomic group.

They are clearly different from the viruses in the genus *Umbravirus* in their genome structure, their molecular biology, and their interactions with plant hosts. However, there is more definitive evidence that they benefit their helper viruses by stimulating helper virus accumulation in co-infected cells. If they stimulate helper luteovirus titer per co-infected cell, the benefits are at least twofold. In co-infections of the ST9aRNA and BWYV, by stimulating BWYV titer more capsid protein molecules are going to be present, thereby ensuring that an adequate supply is available to encapsidate both the progeny BWYV and the ST9aRNA genomic RNAs. Encapsidation of both RNAs is important for survival of both. If the progeny ST9aRNA were to become encapsidated at the expense of encapsidation of the BWYV genomic RNA, this would probably be detrimental to both BWYV and the ST9aRNA. The ST9aRNA requires being in association with BWYV upon subsequent aphid transmission to new host plants. If it is transmitted alone it will probably initiate infections which at best would be limited to only one or a few cells. Also, no BWYV coat protein molecules would be available to encapsidate progeny and it would thus be a dead end with regard to further dissemination. However, by stimulating accumulation of BWYV and the supply of capsid protein molecules, an adequate supply of progeny virions, some containing BWYV genomic RNA and some containing the ST9aRNA, is present and both are likely to be transmitted by aphids together to new host plants. Second, virus titer in the infected host plant for viruses in the family *Luteoviridae* is positively associated with the ability of vector aphids to acquire virus while feeding on the infected plant (GRAY et al. 1991). Thus, a higher titer of both types of virus particles increases the probability that both will be acquired and subsequently transmitted to new host plants.

All of the mixed infections discussed here involve at least one virus which is a member of the family *Luteoviridae*. By comparison, the frequency and different types of virus and/or subviral RNAs associated with luteoviruses is greater than what is seen for other types of plant viruses. Thus, it is tempting to speculate that something about luteovirus biology and/or molecular biology makes these associations possible and that they are likely advantageous. We now have the tools to further study these interactions and attempt to answer these and other questions. It is probable that the near future will be very interesting in regards to increasing our knowledge of viruses and subviral RNAs associated with luteoviruses.

References

Adams AN, Hull R (1972) Tobacco yellow vein, a virus dependent on assistor viruses for its transmission by aphids. Ann Appl Biol 71:135 140

Barker H (1989) Specificity of the effect of sap-transmissible viruses in increasing the accumulation of luteoviruses in co-infected plants. Ann Appl Biol 115:71 78

Chin L-C, Foster JL, Falk BW (1993) The beet western yellows ST9-associated RNA shares homology with carmo-like viruses. Virology 192:473 482

Creamer R, Falk BW (1990) Direct detection of transcapsidated barley yellow dwarf luteoviruses in doubly infected plants. J Gen Virol 71:211 217

D'Arcy CJ, Domier LL, Mayo MA (1998) Family *Luteoviridae*. In: Virus taxonomy. Seventh Report of the International Committee on Taxonomy of Viruses. (in press)

Demler SA, de Zoeten GA (1989) Characterization of a satellite RNA associated with pea enation mosaic virus. J Gen Virol 70:1075 1084

Demler SA, de Zoeten GA (1991) The nucleotide sequence and luteovirus-like nature of RNA 1 of an aphid non-transmissible strain of pea enation mosaic virus. J Gen Virol 72:1819 1834

Demler SA, Rucker DG, de Zoeten GA (1993) The chimeric nature of the genome of pea enation mosaic virus: the independent replication of RNA 2. J Gen Virol 74:1 14

Demler SA, Borkhsenious ON, Rucker DG, de Zoeten GA (1994) Assessment of the autonomy of replicative and structural functions encoded by the luteo-phase of pea enation mosaic virus. J Gen Virol 75:997 1007

Demler SA, de Zoeten GA, Adam G, Harris KF (1996) Pea enation mosaic enamovirus: properties and aphid transmission. In: Harrison BD, Murant AF (eds) The plant viruses, vol 5, Polyhedral virions and bipartite RNA genomes, chap 12. Plenum, New York, pp 303 334

Demler SA, Rucker-Feeney DG, Skaf JS, de Zoeten GA (1997) Expression and suppression of circulative aphid transmission in pea enation mosaic virus. J Gen Virol 78:511–523

Duffus JE (1977) Aphids, viruses and the yellow plague. In: Harris KF, Maramorosch K (eds) Aphids as virus vectors. Academic, New York, pp 361 383

Falk BW, Duffus JE (1981) Epidemiology of persistent helper-dependent aphid-transmitted virus complexes. In: Maramorosch K, KF Harris (eds) Plant diseases and vectors: ecology and epidemiology, chap 5. Academic, New York, pp 162 179

Falk BW, Duffus JE (1984) Identification of small single- and double-stranded RNAs associated with severe symptoms in beet western yellows virus-infected Capsella bursa-pastoris. Phytopathology 74:1224 1229

Falk BW, Duffus JE, Morris TJ (1979) Transmission, host range and serological properties of the viruses causing lettuce speckles disease. Phytopathology 69:612–617

Falk BW, Chin L-S, Duffus JE (1989) Complementary DNA cloning and hybridization analysis of beet western yellows luteovirus RNAs. J Gen Virol 70:1301

Falk BW, Passmore BK, Watson MT, Chin L-S (1995) The specificity and significance of heterologous encapsidation of virus and virus-like RNAs. In: Bills DD, Kung S (eds) Biotechnology and plant protection: viral pathogenesis and disease resistance. pp 391 415

Foster JL (1989) Distribution, variation and biological activity of beet western yellows virus ST9-associated RNA. Masters of Science thesis, University of California at Davis

Gibbs MJ, Cooper JI, Waterhouse PM (1996) The genome organization and affinities of an Australian isolate of carrot mottle umbravirus. Virology 224:310 313

Gildow FE (1987) Virus-membrane interactions involved in circulative transmission of luteoviruses by aphids. Curr Topics Vector Res 4:93 120

Gray SM, Power AG, Smith DM, Seaman AJ, Altman NS (1991) Aphid transmission of barley yellow dwarf virus: acquisition access periods and virus concentration requirements. Phytopathology 81:539 545

Hu JS, Rochow WF, Palukaitis P, Dietert RR (1988) Phenotypic mixing: mechanism of dependent transmission for two related isolates of barley yellow dwarf viruses. Phytopathology 78:1326 1330

Hull RN, Adams AN (1968) Groundnut rosette and its assistor virus. Ann Appl Biol 62:139 145

Miller WA (1994) Luteoviruses. In: Encyclopedia of virology. Academic, New York

Miller WA, Hercus T, Waterhouse PM, Gerlach WL (1991) A satellite RNA of barley yellow dwarf virus contains a novel hammerhead structure in the self-cleavage domain. Virology 183:711 720

Miller WA, Brown CM, Wang S (1997) New punctuation for the genetic code: luteovirus gene expression. Semin Virol 8:3–13

Murant AF (1990) Dependence of groundnut rosette virus on its satellite RNA as well as on groundnut rosette assistor luteovirus for transmission by Aphis craccivora. J Gen Virol 71:2163 2166

Murant AF, Waterhouse PM, Raschke JH, Robinson DJ (1985) Carrot red leaf and carrot mottle virus: observations on the composition of the particles in single and mixed infections. J Gen Virol 66:1575 1579

Nault LR, Gyrisco GG (1966) Relation of the feeding process of the pea aphid to the inoculation of pea enation mosaic virus. Ann Entomol Soc Amer 59:1186–1197

Passmore B, Sanger M, Chin L-S, Falk BW, Bruening G (1993) A subviral replicating RNA stimulates the replication of beet western yellows luteovirus. Proc Natl Acad Sci USA 90:10168 10172

Rasochova L, Passmore BK, Falk BW, Miller WA (1997) The satellite RNA of barley yellow dwarf virus-RPV is supported by beet western yellows in dicotyledonous protoplasts and plants. Virology 231:182–191

Rochow WF (1970) Barley yellow dwarf virus: phenotypic mixing and vector specificity. Science 167:875 878

Rochow WF (1977) Dependent virus transmission from mixed infections. In: Harris KF, Maramorosch K (eds) Aphids as virus vectors. Academic, New York, pp 253–273

Rochow WF (1982) Dependent transmission by aphids of barley yellow dwarf luteoviruses from mixed infections. Phytopathology 72:302–305

Ryabov EV, Oparka KU, Santa Cruz S, Robinson DJ, Taliansky ME (1998) Intracellular location of two groundnut rosette umbravirus proteins delivered by PVX and TMV vectors. Virology 242:303 313

Sanger M, Passmore B, Falk BW, Bruening G, Ding B, Lucas WJ (1994) Symptom severity of beet western yellows virus stain ST9 is conferred by the associated RNA and is not associated with virus release from the phloem. Virology 200:48 55

Schmitz J, Stussi Garaud C, Tacke E, Prufer D, Rohde W (1997) In situ localization of the putative movement protein (pr17) from potato leafroll luteovirus (PLRV) in infected and transgenic potato plants. Virology 235:311 322

Smith KM (1946) Tobacco rosette: a complex virus disease. Parasitiology 37:21–24

Stubbs LL (1948) A new virus of carrots: its transmission, host range, and control. Aust J Sci Res B1:303

Taliansky ME, Robinson DJ, Murant AF (1996) Complete nucleotide sequence and organization of the RNA genome of groundnut rosette umbravirus. J Gen Virol 77:2335 2345

Toros S, Schotman CYL, Peters D (1978) A new approach to measure the LP50 of pea enation mosaic virus in its vector Acytrhosiphon pisum. Virology 90:235–240

Waterhouse PM, Murant AF (1983) Further evidence on the nature of the dependence of carrot mottle virus on carrot red leaf virus for transmission by aphids. Ann Appl Biol 103:455 464

Watson M, Serjeant EP, Lennon EA (1964) Carrot motley dwarf and parsnip mottle viruses. Ann Appl Biol 54:153–166

Watson MT, Falk BW (1994) Ecological and epidemiological factors affecting carrot motley dwarf development in carrots grown in the Salinas Valley of California. Plant Disease 78:477 481

Watson MT, Tian T, Estabrook E, Falk BW (1997) A small RNA resembling the beet western yellows luteovirus ST9-associated RNA is a component of the California carrot motley dwarf complex. Phytopathology 88:164 170

Subject Index

Current Topics in Microbiology and Immunology

Volumes published since 1989 (and still available)

Vol. 217: **Jessberger, Rolf; Lieber, Michael R. (Eds.):** Molecular Analysis of DNA Rearrangements in the Immune System. 1996. 43 figs. IX, 224 pp. ISBN 3-540-61037-5

Vol. 218: **Berns, Kenneth I.; Giraud, Catherine (Eds.):** Adeno-Associated Virus (AAV) Vectors in Gene Therapy. 1996. 38 figs. IX,173 pp. ISBN 3-540-61076-6

Vol. 219: **Gross, Uwe (Ed.):** Toxoplasma gondii. 1996. 31 figs. XI, 274 pp. ISBN 3-540-61300-5

Vol. 220: **Rauscher, Frank J. III; Vogt, Peter K. (Eds.):** Chromosomal Translocations and Oncogenic Transcription Factors. 1997. 28 figs. XI, 166 pp. ISBN 3-540-61402-8

Vol. 221: **Kastan, Michael B. (Ed.):** Genetic Instability and Tumorigenesis. 1997. 12 figs.VII, 180 pp. ISBN 3-540-61518-0

Vol. 222: **Olding, Lars B. (Ed.):** Reproductive Immunology. 1997. 17 figs. XII, 219 pp. ISBN 3-540-61888-0

Vol. 223: **Tracy, S.; Chapman, N. M.; Mahy, B. W. J. (Eds.):** The Coxsackie B Viruses. 1997. 37 figs. VIII, 336 pp. ISBN 3-540-62390-6

Vol. 224: **Potter, Michael; Melchers, Fritz (Eds.):** C-Myc in B-Cell Neoplasia. 1997. 94 figs. XII, 291 pp. ISBN 3-540-62892-4

Vol. 225: **Vogt, Peter K.; Mahan, Michael J. (Eds.):** Bacterial Infection: Close Encounters at the Host Pathogen Interface. 1998. 15 figs. IX, 169 pp. ISBN 3-540-63260-3

Vol. 226: **Koprowski, Hilary; Weiner, David B. (Eds.):** DNA Vaccination/Genetic Vaccination. 1998. 31 figs. XVIII, 198 pp. ISBN 3-540-63392-8

Vol. 227: **Vogt, Peter K.; Reed, Steven I. (Eds.):** Cyclin Dependent Kinase (CDK) Inhibitors. 1998. 15 figs. XII, 169 pp. ISBN 3-540-63429-0

Vol. 228: **Pawson, Anthony I. (Ed.):** Protein Modules in Signal Transduction. 1998. 42 figs. IX, 368 pp. ISBN 3-540-63396-0

Vol. 229: **Kelsoe, Garnett; Flajnik, Martin (Eds.):** Somatic Diversification of Immune Responses. 1998. 38 figs. IX, 221 pp. ISBN 3-540-63608-0

Vol. 230: **Kärre, Klas; Colonna, Marco (Eds.):** Specificity, Function, and Development of NK Cells. 1998. 22 figs. IX, 248 pp. ISBN 3-540-63941-1

Vol. 231: **Holzmann, Bernhard; Wagner, Hermann (Eds.):** Leukocyte Integrins in the Immune System and Malignant Disease. 1998. 40 figs. XIII, 189 pp. ISBN 3-540-63609-9

Vol. 232: **Whitton, J. Lindsay (Ed.):** Antigen Presentation. 1998. 11 figs. IX, 244 pp. ISBN 3-540-63813-X

Vol. 233/I: **Tyler, Kenneth L.; Oldstone, Michael B. A. (Eds.):** Reoviruses I. 1998. 29 figs. XVIII, 223 pp. ISBN 3-540-63946-2

Vol. 233/II: **Tyler, Kenneth L.; Oldstone, Michael B. A. (Eds.):** Reoviruses II. 1998. 45 figs. XVI, 187 pp. ISBN 3-540-63947-0

Vol. 234: **Frankel, Arthur E. (Ed.):** Clinical Applications of Immunotoxins. 1999. 16 figs. IX, 122 pp. ISBN 3-540-64097-5

Vol. 235: **Klenk, Hans-Dieter (Ed.):** Marburg and Ebola Viruses. 1999. 34 figs. XI, 225 pp. ISBN 3-540-64729-5

Vol. 236: **Kraehenbuhl, Jean-Pierre; Neutra, Marian R. (Eds.):** Defense of Mucosal Surfaces: Pathogenesis, Immunity and Vaccines. 1999. 30 figs. IX, 296 pp. ISBN 3-540-64730-9

Vol. 237: **Claesson-Welsh, Lena (Ed.):** Vascular Growth Factors and Angiogenesis. 1999. 36 figs. X, 189 pp. ISBN 3-540-64731-7

Vol. 238: **Coffman, Robert L.; Romagnani, Sergio (Eds.):** Redirection of Th1 and Th2 Responses. 1999. 6 figs. IX, 148 pp. ISBN 3-540-65048-2

Springer
and the
environment

At Springer we firmly believe that an
international science publisher has a
special obligation to the environment,
and our corporate policies consistently
reflect this conviction.
We also expect our business partners –
paper mills, printers, packaging
manufacturers, etc. – to commit
themselves to using materials and
production processes that do not harm
the environment. The paper in this
book is made from low- or no-chlorine
pulp and is acid free, in conformance
with international standards for paper
permanency.

Springer